VIDEO
CASSETTE RECORDERS
BUYING, USING & MAINTAINING
BY BILL PASTERNAK

TAB BOOKS Inc.
BLUE RIDGE SUMMIT, PA. 17214

8307505

To my good friend Dave Bell. An artist of film and television production in his own right, he has been a true inspiration for me since we met almost ten years ago. From our friendship I have learned many things, including that there is no substitute for excellence in any undertaking. This is my thank you for a friendship I know will last a lifetime.

FIRST EDITION

FIRST PRINTING

Copyright © 1983 by TAB BOOKS Inc.

Printed in the United States of America

Library of Congress Cataloging in Publication Data

Pasternak, Bill.
    Video cassette recorders.

    Includes index.
    1. Video tape recorders and recording—
Amateurs' manuals. I. Title.
TK9961.P37 1983   621.388′33   82-19397
ISBN 0-8306-0490-1
ISBN 0-8306-1490-7 (pbk.)

Contents

Foreword   vii

Acknowledgments   xi

1 Introduction to Video Recording   1

2 Evolution of Video Recording   4
Fast Scan TV—Early Video Recording Developments—Modern Equipment

3 The First Videotape Recorder (VTR)   10

4 The Video Recording Process   14
Basic Principles—The Vidicon Tube

5 Some Made It, Others Didn't   21
U-Matic—Betamax (Beta)—Video Home System (VHS)—Losers

6 Betamax versus VHS   25
Basic Betamax (Sony Model SLO-320)—Basic VHS (General Electric Model 1VCR2002X)—Tape Handling

7 The Video Head   36

8 The Complete Video Cassette Recorder (VCR) System   40
Record/Playback Deck—System Control Unit—Signal-Processing Package—Audio Treatment

9   How to Buy a VCR   49

10 Getting the Most Out of Your VCR   53
    The TV Receiver and Antenna—VCR Maintenance—Tape
    Techniques—Video Cameras, VCR Heads, and VCR Ac-
    cessories—Video Editing—Audio and Microphones—Film
    Transfer to Videotape—Taping From Cable TV and Interference

11 The Video Camera   75
    Professional Standards—How to Compare Home Video Cameras

12 Video Camera Do's, Don'ts, Questions, and Answers   86
    A Dozen Do's—A Baker's Dozen Don'ts—Questions and
    Answers about Video Cameras

13 Your TV Set, the Last Link   94

14 Video Software   102
    Video Clubs—Blank Videotape—Pre-recorded Software

15 Making a Video Movie   109
    Concept Development—The Outline—Equipment Selection—
    Scripting—Location and Trial Shoot—Evaluation and Post-
    production Editing

16 The Legal Aspects of Home Video Recording   122

17 Videodisc Systems   126

    Appendixes: Video Movie Documentation for
                 "You-Me and Energy"

    A Concept   131
    B Equipment List   134
    C Script   136

    Index   139

Foreword

Videotape recording has had a profound impact on man's ability to exchange thoughts and ideas, to learn, work, be entertained, and play since Ampex Corporation developed the technology 25 years ago.

In fact, it is safe to say that the average American cannot go through a day without encountering videotape recording in one form or another. And that includes an astronaut who may be journeying to the moon.

Videotape recording and playback has also permitted the creation of a daily chronicle of present and recent history for large worldwide audiences, from the drama of man on the moon and the resignation of a President to the tension of a Super Bowl battle.

The playback of these events on video tape can be almost instantaneous—and has led to a familiar phenomenon, the "instant replay" of history.

While its effect on broadcasting has been substantial (more than 85 percent of all programming is broadcast from videotape recordings), video tape has brought new techniques to other fields.

In medicine, many hospitals are taping operations for later study by surgeons and students. Instant replay itself is being widely utilized to permit doctors to check procedures during an operation.

Educators videotape lectures and demonstrations for repeated usage and for distribution to schools and universities throughout the world.

There wouldn't have been pictures of man in space without videotape. Analog signals from the moon or spacecraft transmitted to earth were processed and translated by instrumentation equipment into electronic impulses for videotape.

The lives of many sports officials would be less complicated if the instant replay wasn't watching every play, but then many golf and tennis swings would remain uncorrected because there was no videotape recording to show what was wrong.

Business utilizes videotape in information storage as an adjunct to the computer, in security surveillance of company facilities, in operating procedures, safety instruction, and other activities.

U.S. Navy aircraft carriers record on video tape planes taking off and landing on their decks as a training aid and safety procedure.

Live television newscasts use videotape heavily, taping earlier news events of the day for later showing.

And, of course, millions of homes around the world are now equipped with their own videotape recorders for taping and viewing movies and other programming.

Videotape recording has come a long way since April 14, 1956, when a team of engineers from Ampex first publicly demonstrated a practical videotape recorder (VTR). Then basically a time delay device, the videotape recorder has been refined and improved to a point where it is now a sophisticated communications system that can record, edit, add special effects, and even improve the quality of the picture before it is replayed.

These advances have been accompanied by a reduction in size and weight, from bulky machines weighing well over a ton to portable professional VTRs—such as the Ampex VPR-20—that weigh just 53 pounds.

Ampex has remained on the leading edge of these changes and is still the world's largest manufacturer of professional videotape recorders. The company has sold more than 60 percent of the over 20,000 professional VTRs that have been delivered in the last 25 years, and Ampex licenses its technology to all manufacturers of home videotape recorder systems, although it makes no consumer recorders itself.

The videotape recording procedures developed by Ampex and accepted as the standard by government and broadcasters involves the recording of pictures converted to an electrical current by a television camera much in the same way sound is converted to an electrical current by a microphone.

Magnetic tape, composed of a plastic base coated with minute particles of metallic oxide, moves past magnetic recording heads that create fields within the tape and align the particles in specific patterns. For replay, the tape passes a replay head and the previously aligned patterns disturb the field of the head to cause generation of electrical signals that result in pictures and sound.

Recording heads are electrically energized magnets. A picture to be recorded is transmitted to the head in the form of electrical signals from a television camera. Variations in the signals cause the head magnet to vary the field it produces and align the oxide particles on the tape.

Magnetic tape recordings can be used repeatedly and stored for long periods without significant degradation in quality. Editing is also simple and economical. Unlike film, where segments must be cut and glued, videotape is edited electronically. Unwanted information can be erased by demagnetizing so the tape can be used repeatedly. Videotaping permits an immediate look at the action: there's no waiting for film processing.

And what of the future? Videotape is helping to create new communications systems in emerging nations around the world, a major and growing market for Ampex.

In law, videotapes of bank robberies and other crimes are being shown to juries. On some states, even court proceedings are being videotaped.

Business leaders are finding videotape an effective means of communicating with their employees, shareholders and customers. That activity is growing rapidly.

It even has the potential of providing an extra official at sports events through the use of instant replay.

The Ampex Corporation

Acknowledgments

I gratefully acknowledge the assistance of the following individuals and companies for their help in the preparation of this book:

Ampex Audio-Video Systems
3-M Corporation
The Magnavox Corporation
Mitsubishi Electric Co, Ltd.
Sony Corporation: Broadcast Video and Consumer Electronics Divisions
Panasonic Video Systems
Panasonic Consumer Products
The Video Services Group, Canyon Country California
The General Electric Co: Consumer Video Products
Caston's T.V. and Appliance Co, Newhall Ca.
Ikegami Electronics, USA
Hitachi Corporation: Broadcast Video and Consumer Electronics Divisions
Sanyo Electric Corp: Consumer Products Division (Janek Skok and Russell Mayworm)

Mr. Lee Scott
Mr. Jerry Caston
Mr. Rupert Goodspeed
Mr. Albert Benson, Metromedia Corp. Los Angeles

With a very special word of thanks to Mr. George Boardman and the People at Ampex for their historical material on the development of video recording. Without their dedication, video recording and this book might never have come about.

Chapter 1

Introduction to Video Recording

Not since the introduction of television in 1940 has any product in consumer electronics so captured the imagination of the buying public as the video cassette recorder. Only a few years ago, a video recorder was an expensive luxury enjoyed by a select few. Then with Sony's introduction of the Betamax recorder a whole new era in home entertainment was upon us. Sony's Betamax was an easy-to-use video recording system at a price most consumers could afford. The first series of machines had some disadvantages such as a short (by U.S. standards) recording time period, if you wanted the luxury of delayed automatic recording, you had to purchase extra components. The picture quality, however, was good, and audio was comparable to what the television set itself could reproduce. The Betamax soon became a household word in this country and elsewhere.

Before the Betamax and *Beta format* of recording there had been other entries into the consumer market. In fact, my first exposure to what might be considered "home recording of video" came through a friend named Henry Feinberg. Back in the late 1960s Henry purchased a Sony AV-3000 black-and-white portable video recorder. His "toy" was the state of the art videotape recorder (VTR) at the time, even though it needed some massive interconnections to the television (TV) set in order to operate. Henry and I modified his RCA CTC 5 series color set to provide both audio and video signals for the VTR. Luckily, the machine did have the ability to play back from the TV by connecting it to the antenna terminals. The results rather amazed us. Quality was about equal to Super 8 motion picture film.

This machine was not intended for use in a broadcast station. It was termed an "industrial/educational VTR" and did not even use cassettes. Its

1

configuration was that of open reel-to-reel, with a forward tape speed of 7½ inches per second (IPS). This gave a maximum one-hour recording on a 7 inch reel of tape. The tape width was ½ inch, the same standard used in home VCRs today. For what it's worth, this very machine is still around. In fact I now own it, and it's not collecting dust in the closet. When not in use as part of my ham television station, it's on loan to someone who has old tapes and wants to transfer them to a more modern format.

Not long after the Betamax hit the U.S. marketplace, another home recording system emerged and was promoted very heavily by two of Sony's chief competitors: Panasonic and RCA. This Video Home System (VHS) has also withstood the test of time to become one of the two accepted home entertainment standards. Because its recording time is longer than the beta format, it soon gained wide acceptance in the consumer marketplace. For a while, many thought the beta format would die a very quick death, but that was not the case. The recording time span on Betamax was upped to almost that of VHS, and Betamax was the first to offer specialized "bells and whistles" that the U.S. consumer was begging for. Both formats have survived, both are accepted and both can trace their lineage to yet another cartridge tape format: the "big brother" and standard of electronic TV journalism, the ¾-inch U-Matic. While the manufacturers may not wish to admit it, any competent VCR/VTR technician will tell you that the similarities between half-inch beta and VHS formats and three-quarter-inch commercial machines have to be a lot more than pure coincidence. All you need to do is look inside to realize that the half-inch home machines are almost identical to their ¾-inch ancestors. In fact, the first time I took a peek inside the early Sony SL-7200 Betamax machines, I thought I was viewing a shrunken version of my VO-1600 U-Matic. It was obvious that the same engineering was common to both.

These days I make my living from videotape and video recording. I work for one of the largest TV production organizations in the nation whose backbone is video recording. The "goodies" I have access to would make the true videophile's mouth water. The smallest format used for commercial broadcasting is the ¾-inch U-Matic, with most production being performed on 1-inch helical scan. Another one-inch tape format, called "type C" in the broadcast industry, that in a few short years has made 2-inch quadruplex, a videotape developed by Ampex Corp. and used in broadcasting, all but obsolete. For years, 2-inch videotape had been the industry standard, and it will probably be around for a while yet. We have other magic devices like the Chiron used for putting titles on the screen, squeeze-zoom, and chroma key.

Maybe one day such formats will be part of every home video recording setup. Right now they are out of the price range of all but the richest of videophiles. They are there, but way beyond the scope of this book. What I am interested in doing is introducing you to a bit of the technical side of that "magic box" you already own and showing you what visions can come from it without breaking the bank. You won't become a video service technician from reading what is yet to come, but you will understand your VCR, it's abilities, and its limitations.

Chapter 2

Evolution of Video Recording

If I told you that you could record and replay color television images on a $20 cassette recorder, you would probably say one of two things. Either the guy writing the book is crazy, or he knows something the rest of us don't. Well, I am not crazy. Ham radio operators have been sending television images worldwide for years. And yes, many of them record, store, and replay these images on "five-and-dime" recorders. You could do it as well if you held the necessary license, owned some other expensive equipment, and did not mind the limitation of viewing a still picture for eight seconds at a clip. The system, called slow-scan television (SSTV), was developed in the late 1950s by Copthorne MacDonald while he was an engineering student at the University of Kentucky.

FAST SCAN TV

If you are an average consumer, not a ham radio operator, the thing that will probably interest you the most is being able to record your favorite TV show while you are away from home. This is called time-shift recording, and just about every VCR on the market has some provision for doing this.

The system used for live television, regardless if "live" means happening that moment or on taped delay is called *fast scan* (FSTV). In the United States, Canada, Mexico, parts of South America, Japan, and a few other spots, the adopted system for transmitting a video signal with color (or chroma) information is called NTSC (National Television Systems Committee). Two other color video systems used in other parts of the world are called PAL and SECAM. NTSC is based on 525 lines of picture or "resolution" while both PAL and SECAM use 625 lines as their base. Yes, the 625 line system is superior to 525 lines, but we are locked into the

NTSC signal standard for economic reasons and it's doubtful that there will ever be any major change. It's interesting to note that in some parts of the world, the TV sets and video recorders sold are of multi- or tristandard design. This is because neighboring nations with overlapping transmission coverage may use different transmitting systems. England uses PAL, and France, right across the English Channel, uses SECAM. Here, the reason for multistandard receivers and VCRs is obvious. Actually, there are more than just these three TV systems around, but they account for most of the TVs and VCRs in the world.

If you are interested in getting more information on TV standards from around the world, obtain a copy of the *World Radio & TV Handbook*. This is an annually updated anthology of all known broadcast activities worldwide, and in itself is a truly international undertaking. The handbook is written in Denmark, printed in England, and distributed worldwide. This volume is available wherever books on radio and communication are sold and is considered the "bible" of the industry.

EARLY VIDEO RECORDING DEVELOPMENTS

I have not yet answered the simple question raised earlier: why can't you record television pictures at "full speed" (the term for this is normal scan rate) on your audio recorder? There is a reason and it's called bandwidth. Let's look at the specifications of the normal "dime store" cassette recorder. Most can record and play up to two hours on a type C-120 cassette. That's good. Most TV movies and sports events are about that long (with the commercials removed). Let's see, they have automatic record level control. That's good. Just pop in a cassette and push a button; no knobs to fiddle with. But wait, what's this? Frequency response is from 80 Hz to 10,000 Hz. What does that mean? Simply, it means that the recorder will respond to signals in the frequency range specified and has the ability to reproduce those same signals. This range of signals is called "bandwidth," and a TV signal is 4-megahertz wide, depending on where you live in the world. That little $19.95 recorder just "can't hear" that frequency range.

If my little cassette machine can't record pictures, why then can a machine be made that will? How in fact can it be made? The story goes back to the 1950s when a number of companies realized that an all-electronic method of retrieving recorded information was becoming essential to the industry. Most of the major broadcast electronics firms began development programs for such a system. The first on the scene, and still considered the "grand-daddy of them all," was a unit developed by the Ampex Corporation. It used 2-inch-wide tape and a rotating head assembly that recorded across the width of the tape. The system was dubbed Quadruplex, and with the advent of color it went through an evolution process where it first entered the "low-band" era and finally the "high-band" era, where it remains today.

In time, tape began replacing film for in-studio work, but for news coverage, film was still the front runner. Electronic cameras were massive in size and required a myriad of support electronics and personnel. Tape machines? A single unit could fill half a room, not to mention the need for a generator truck to run the entire mess. Except for certain circumstances, videotape would stay in the studio and 16mm film would carry on in the field for news.

In the '60s, however, the three major TV networks began building massive mobile units for field production. Most were multiple 50-foot tractor-trailer vehicles with the generator truck following close behind. There was very little attempt at miniaturizing. Let's face it, a tape machine that was ten feet wide in the studio would take the same ten feet in a trailer. Also, in that era much of the equipment was vacuum tube type, making the early electronic packages much larger than their solidstate offspring of today. One of the most famous early TV production units was called "RED-EO-TAPE" (built for Red Skelton). It too was a massive multi-vehicle affair (see Fig. 2-1), through totally state of the art for its day. One of it's designers was Rupert Goodspeed and Rupe was awarded an Emmy for his work.

As this book is being written, Rupert and I have joined forces to build another such vehicle, the Mobile Production Unit (MPU). The MPU had to do everything its predecessor accomplished and more, yet be no larger than a Ford SuperVan. In fact, the unit would be built into such a vehicle. By the time this book reaches publication, the MPU will have been in service

Fig. 2-1. This is RED-EO-TAPE, one of the nation's first mobile videotape production units. Today, a single van can do the same job it used to take three semi-trailers to do. This installation won an Emmy! (Courtesy Rupert Goodspeed.)

several months, going five or six days a week in production of the first commercial cable TV program. It will carry a pair of color cameras, a pair of video recorders, and everything else necessary to produce a TV series in the field. The two videotape recorders it will carry (one a 1-inch reel-to-reel and the other a ¾-inch U-Matic) are both about the same size as the early home recorders.

MODERN EQUIPMENT

I mentioned the MPU to show how times have changed and to illustrate a definite evolution in video recording technology. For example, a studio console 1-inch helical machine is about one-fifth the size of its 2-inch predecessor. Field cameras weigh from 20 to 30 lbs and can easily be hoisted onto a cameraman's shoulder, the same as a film camera. A ¾-inch field recorder is not that much larger than its ½-inch home-entertainment cousin and can easily be strapped onto the cameraman's back for one-man operation. The equipment used by the broadcaster costs a lot more than your home Beta or VHS unit because of tighter specifications and more rugged design. A 1-inch portable recorder costs in the neighborhood of $30,000-$40,000 depending on make and features included. Mini-cams, as the video cameras are called, cost from $12,000 up to $80,000. Compare that to a complete camera/recorder setup for home use selling for under $2,000.

The two classes of equipment have something very much in common. Both use a recording system called "helical scan," and it's this system that has made the miniaturized equipment and lower cost in the consumer market a fact of today. Helical scan is not anything new, since the Sony black-and-white recorder (monochrome) used this very same principle. It's just that the engineers have refined it a lot more, making possible the use of sealed tape cassettes, color recording, and greater time capabilities. The big breakthrough came in the late 1960's when Sony introduced the ¾-inch U-Matic format for broadcast, educational, and industrial use. It featured up to an hour recording in full color, excellent recovered video/chroma resolution, instant loading from a sealed tape cassette, and multiple audio tracks. These machines soon started showing up among the true videophiles. They were (and still are) far from inexpensive.

A short time back, I mentioned that the NTSC standard color system uses 525 lines. What I did not say at the time was that few of the TV sets we have in our homes show us that many. If we are lucky, we get to view somewhere between 250 and 300 lines of video information. Normal transmission losses, poor reception, and TV set design tend to keep the actual number of recovered lines below 300. Also, some of the transmitted lines do not contain information of interest to the TV viewer. Have you ever "rolled" the TV picture up and down with the vertical hold control and noticed a black area with some white bars. This is information transmitted by the TV station to control certain circuits in your television. Without it, your picture would roll and tear from side to side, up and down. These bars are called synchronizing signals or simply "sync." Also in that area are VIR (a color control standards signal) and closed captioning for the hearing-impaired viewer. While your TV set and some viewers definitely benefit from this portion of the transmitted signal, obviously it's something that you will not bother to watch.

It becomes evident that if a video recorder can accurately reproduce even 250-300 lines of usable video information, the average viewer will not be able to tell the difference between it and the actual broadcast. Many home recorders can reproduce a better image than the TV set. To realize the full capabilities of a video recorder requires that it be viewed on a studio-quality monitor, which cost from $4000 to $7000. They may give the very best picture that money can buy, but they don't exactly fit with most living room decor. Even the metal cases for these gems are an option that won't lend elegance unless your living room is supposed to look like "master control." While every recorder made for the consumer has video and audio output jacks for such a monitor, they are seldom used. Most of those who purchase a video cassette recorder simply hook it to the antenna terminals on their TV sets and RF-feed channel 4 or 3. More losses.

Let's say that the average home recorder can reproduce up to 330 lines. I say average because of my three home-format VCRs, one exceeds that number, one just hits it, and the last just makes 270 lines. (These results were obtained using a pre-recorded test tape and feeding the recorder's video output into a test setup consisting of a Videotek waveform monitor, RCA vectorscope, and Ikegami Studio type 17″ line monitor. The test tapes were recommended by the manufacturers and the recovered video resolution level was read from the monoscope patterns on the test tapes. Viewing off-air recorded tapes on any of the three recorders on a Sears/Sanyo 17-inch portable color TV set, it's almost impossible to see any difference. What becomes evident is that if a home video recorder can slightly exceed what the average viewer is used to seeing on his one-eyed monster, then the reproduced signal from the VCR will look good to that "average" viewer.

An excellent example is my own ¾-inch U-Matic which runs rings around any of my ½-inch machines on both the test fixture and when driving my Sony TV converted to direct video/audio drive. It looks as though you were in the control room in that configuration, but into either of my other sets through the antenna terminals it's no better a picture than any of the ½-inch machines. This proves two things: Most TV viewers are seeing far less in recovered video than they think, and most VCRs have the ability to provide even a better picture than their owners expect. Don't get upset. These are "average" conditions and do differ from home-to-home.

The evolution of the home recorder follows a definite line. To under-stand where it all started, you must accept the simple truth that the home VCRs of today are with us because there was a need elsewhere to develop such items. Even the "bells and whistles" such as high-speed search, freeze-frame (pause with video on the TV screen), and low motion were developments for the broadcaster that were eventually engineered into the home video cassette recorder, as surveys by the manufacturers showed that the public wanted such goodies.

What will the future hold? Your guess is as good as mine. I say this because in doing research for this book, I wrote to almost all manufacturers of home entertainment video recording equipment. While most were happy to provide material about their current product line, only one such organization, Mitsubishi Electric, was kind enough to permit a glance into their soon-to-be-released consumer video products. (This is not meant as "sour grapes".New product security is a paramount concern, and to those companies who could not provide advance information I can only say that I understand their situation.)

By now, you should have some idea of what television and video recording are, and I'll bet it did not hurt one bit. As long as we are on a high "roll" as they say in Las Vegas, let's move on to the next chapter to get a bit deeper into the concepts outlined thus far, and see if we can make a bit more sense out of the whole video recording process.

Chapter 3

The First Videotape Recorder (VTR)

The following information was supplied by the Ampex Corporation, Redwood City, California, and is reprinted with their permission. I gratefully acknowledge the kind assistance of Mr. George Boardman. Media Relations Manager of Ampex in obtaining this and other material presented in this book.

Videotape recording is so widespread and commonplace today, it is difficult to believe the technology astounded television broadcasters and created a multibillion dollar industry just 25 years ago. But in April 1956, the ability to broadcast from tape stirred pandemonium among the viewing audience and revolutionized the television industry. The event climaxed four years of work by a six-man team at Ampex Corporation that would not give up its dream of recording television pictures on magnetic tape.

The story of how that dream came true started in 1951 with the arrival of Charles P. Ginsburg at Ampex. Ginsburg, now vice president—advanced technology planning, joined the Redwood City, California, firm for the express purpose of putting pictures on tape.

At the time of Ginsburg's arrival, Ampex made professional audio tape recorders for radio broadcasters and recording studios. But it didn't take much business acumen in the early '50s to realize that recording pictures on tape with the same ease and versatility with which sound was being done would lead to a large pot of gold at the end of the rainbow.

While the prize was obvious, the path to it was strewn with so many technical roadblocks that only the rich or foolhardy could pursue it. Two affluent organizations—RCA in the U.S. and the BBC in England—were committing skilled staff and healthy budgets to developing high speed videotape recorders (VTRs) if only to prove that pictures could be put on tape. The machines worked to some degree, but they gobbled up thousands of feet of tape in a few minutes and were commercially impractical.

While Ginsburg and his team were quietly experimenting with rotary head principles (the tape moves slowly and a magnetic head revolves rapidly), singer Bing Crosby also recognized the financial potential and set up his own experimental group in Hollywood. Fortunately for Ampex, this well equipped and amply funded group also went the route of fixed heads and high speed tape, a blind alley that was eventually to be abandoned by everyone when the first Ampex videotape recorder was introduced in 1956.

What was so difficult about developing a VTR? Surrounded now by a gamut of tape machines of all sizes and shapes, it seems in retrospect that it must have been easier than it was. In the late '40s, high fidelity audio tape recording was still a technical achievement of some novelty. While our ears need only 15,000 cycles per second to hear satisfactorily, our eyes demand at least 4 million cycles per second for acceptable pictures and no known magnetic mechanism of that era could even come close to that figure.

Ampex explored the idea of departing from narrow (¼-inch wide) tape moving over stationary magnetic heads and trying wider (2-inch) tape moving relatively slowly (15 inches per second) past rapidly revolving heads. After some early experiments that were disappointing, Ginsburg was fortunate to encounter a part-time Ampex employee and college student, Ray Dolby, whose interest and dedication to the early video recording attempts led to some measure of success.

By November of 1952, Alexander M. Poniatoff, the company founder, thought it appropriate to issue a memo attesting to a witnessed video tape playback whose spectators were himself, Ginsburg, Dolby and the company's patent attorney. He commented in the memo that this was being done in case of future patent problems, a prophetic statement in view of the subsequent scramble to patent other VTRs around Ampex's original creation.

Ampex was neither large nor wealthy enough at that time to assign unlimited priority to what was then called the TVR project, but that didn't prevent Ginsburg and Dolby from living and breathing the dream of committing pictures to tape. In 1953 Ginsburg convinced management to spend a few more precious hours and dollars. The results were promising enough to increase the development team to six people (see Fig. 3-1), including engineers Charles Anderson, Alex Maxey, and Fred Pfost. The final member was Shelby Henderson, a machinist who built and designed the mechanical components.

The group faced some difficult problems. In 1954, no tape manufacturer made a suitable tape, no magnetic heads worked at the desired frequencies, no signal systems could cope with the fluctuations from poor tape, and no motors rotated smoothly enough to make stable images. The list of secondary problems was endless, and most agonizing was the interdependency of the solutions; curing one thing often upset something else.

Fig. 3-1. The six men who developed the first practical videotape recorder at Ampex Corporation are shown with the result of their work, the Mark III, and the Emmy Award won by Ampex for the achievement. They are, from the left, Charles E. Anderson, Ray Dolby, Alex Maxey, Shelby Henderson, Charles Ginsburg, and Fred Pfost.

The task was split up so that each man worked on his area of interest and expertise.

While there was a high team spirit and endless effort toward their mutual goal, there were also some low psychological valleys when announcements of imminent success came from RCA or the Crosby organization. But the group maintained its optimism by reading the fine print in the news releases; there were still a few minor technical details to clean up, like getting more than a few minutes of playing time from a reel of tape three feet in diameter.

Ginsburg is credited by his colleagues for not only the concept but, more importantly, for maintaining the momentum of the project in the face of seemingly unattainable objectives. Each team member worked with patience and precision to break the bottlenecks: switching, the FM signal system idea, the miniscule magnetic head assemblies, and the dynamics of tape scanning to build a basic system that has held for 25 years.

The fruits of their labors became apparent April 14, 1956, when the first videotape recorder was demonstrated at a meeting of 200 CBS affiliates in Chicago. The audience of TV engineers and station managers, realizing they were witnessing history, stood up, stamped, cheered and whistled their approval. The National Association of Radio and Television

Broadcasters convention followed that demonstration, and Ampex left the convention with orders for $4 million worth of its magnetic miracle.

Today, the worldwide broadcast industry has absorbed well over $1 billion worth of high quality VTRs and accessories for TV program origination, sophisticated editing, slow and stop or reverse action, and all the other unique effects that are daily fare on the airwaves.

Ampex has built and sold the major portion of these VTRs and in the process has grown into a $500 million a year corporation with nearly 13,000 employees and offices in every corner of the video world.

Chapter 4

The Video Recording Process

To understand how a video recorder and player works, it is first necessary to explain the recording and retrieval process itself. There are many texts on this, mainly for use by engineers and maintenance technicians. They are filled with highfalutin mathematical equations and equally tough terminology, not exactly your Sunday afternoon fun reading material. Nevertheless, knowing the basics is important, and in this chapter with the aid of some diagrams and outlines I will try to explain things in a simple way.

BASIC PRINCIPLES

Let's start by looking again at our $19.95 audio cassette machine. How does it work? When you talk into the microphone as shown in Fig. 4-1, your voice hits a unit called a microphone. All the microphone does is convert your voice sound to electric impulses which are a direct representation of what you said. I am not going to explain how a microphone works, but rather we will assume that it has the ability just described.

The sound waves now converted to electric impulses are connected to a part of the recorder called an audio amplifier. The amplifier is nothing more than an electronic circuit designed to make the signal coming from the microphone larger in level or "amplitude." In most simple cassette voice recorders this is accomplished by three or four transistors or a single integrated circuit chip. Either way, the output signal level of the amplifier is large enough to make the *record head* work. What is the record head? Nothing more than a tiny electromagnet, a pint-sized version of the big electromagnets you will remember from high school science classes. To make an electromagnet work, you pass an electric current through a coil of wire which is wrapped around an iron (or other ferrous metal) core. The

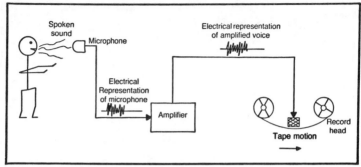

Fig. 4-1. Basic audio tape recording.

simplest electromagnets are made by winding dime store hookup wire around a large nail, and then connecting the wire ends across a flashlight battery. Voila! An instant electromagnet (see Fig. 4-2).

Now let's suppose that you put a switch in series with one of the wires from your homemade electromagnet to its battery. Each time you closed the switch the magnet would be turned on. Conversely, when you opened the switch the magnet would be turned off. If you began to turn the switch on and off very quickly, you would have an electromagnet that would be picking things up and then dropping them. For example, if there were a box of pins right under your electromagnet, it would pick them up and drop them, pick them up and drop them, pick them up and drop them, etc. This would continue as long as you kept opening and closing the switch.

Now, let's substitute an ac source for the battery (which is dc). In ac, or *alternating current* electricity, the voltage continually goes from zero to a positive peak then back to zero and then to a negative peak. (See Fig. 4-3). We now have connected our electromagnet to an alternating current source

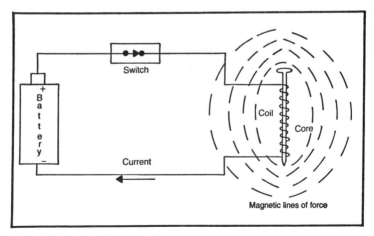

Fig. 4-2. Simple electromagnet.

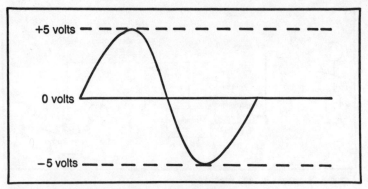

Fig. 4-3. Electrical representation of ac electricity.

that supplies the same peak voltage to our electromagnet, but slowly alternates on and off. What would happen when you hold the electromagnet over the box of pins now? Exactly the same thing as when you opened and closed the switch. The pins would be picked up and dropped once every second or so. If we raised the frequency of the ac to the normal 60 Hz (cycles per second) found in your house, the pins would seem to stick to the magnet the same as when you were connected to the battery with the switch turned on. In reality, the magnet is letting go of the pins sixty times each second, but before the pins can drop away it's turned on again so the pins appear to be held firmly.

What does all this have to do with recording a sound or a picture? Electromagnets and alternating current electricity are the basis of all magnetic recording. To understand, let's go back to Fig. 4-1 and note the little electrical symbols in it. These are called wave diagrams, and you will note that they are ac waves. In fact, they look a lot like Fig. 4-3. This is because the electricity produced by a microphone and the electricity made bigger by the amplifier is ac. The microphone converts your voice to ac electricity and feeds it to the amplifier which increases its level to a point where it can operate the recording head, a tiny electromagnet. Recording tape is nothing more than a strip of plastic coated with a ferrous material. As the tape passes by the recording head, molecules of ferrous material are lined up to correspond with the amount of magnetism the head is putting out at a given instant. To explain the exact nature of this process requires some exotic mathematics which are beyond the scope of this book. Just assume that the process works as I have described, and what you said into the microphone is instantly recorded on the tape.

To play back the recording, we use another basic scientific principle—that of the electric generator. (See Fig. 4-4). Here, the tape, containing ferrous material, is passed near a coil of wire wrapped around a core. Movement of the tape past this *playback head* creates a magnetic field which in turn induces voltage in the coil of wire. Because the magnetic field is continually changing, an ac voltage is developed. Like the microphone, the

16

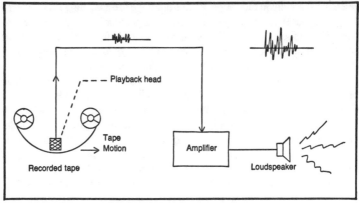

Fig. 4-4. Basic audio tape playback.

voltage produced by the playback head is quite small (only a fraction of a volt) and must be made larger to drive a loudspeaker. So, the output of the playback head is fed into an amplifier and the output of the amplifier is connected to a speaker. Turn on the mechanism that moves the tape forward, and your voice comes forth. This is a capsule version of how magnetic recording and reproduction takes place. Now, let's apply this to video recording.

THE VIDICON TUBE

How do you convert pictures to electric impulses? Pictures are converted to electricity by a device known as a vidicon tube. There are many variations on the same basic principle, but if you gain an understanding of how a vidicon tube works, you will be able to understand more complex devices (such as tri-electrode devices for single-tube color cameras).(see Fig. 4-5). When light hits the photoconductive target element of the vidicon tube, it causes electrical resistance to change in relation to the amount of light hitting the particular spot on the target. A dc voltage is fed to the target through a load resistor (RL). This voltage creates an electrical charge which is added to the vidicon's internal resistive/capacitive structure. An electron beam from the other end of the vidicon scans the target and causes a varying ac voltage to be developed across RL. The voltage is in direct instantaneous proportion to the resistance of a given location on the target and the resistance of that spot (called a picture element) based upon the amount of light hitting it. Kind of complex you say. Okay, let's just say that a vidicon is a device that turns light into electrical impulses usable for video purposes, akin to what a microphone does for sound waves. Also, like a film camera, if you place a lens in front of the vidicon tube, you can focus the image that the vidicon sees. If you then connect the output of the vidicon through an amplifier to the screen of a TV monitor (or through the combination of an amplifier and RF modulator to the terminals of a TV set) you will see what the vidicon tube sees. At this point we are assuming FSTV at 525 lines, and that the necessary electronic circuits are connected to the rest of the vidicon tube elements. The physical structure of a typical vidicon tube can be seen in Fig. 4-5. It's very complex and quite expensive, but it's also the "heart" of converting images to electronic video.

At this point we have our video camera and we want to record our pictures on videotape. How is this done? While the technology involved differs from audio recording, the basic principle is the same. The video informatin is amplified and fed to a recording head assembly in a videotape recorder is called a *scanner*. In audio recorders, the record head is stationary while in video recorders the head rotates at a very precise speed. Rotation is necessary in order to fit all the video information onto the tape. Remember, we are recording a signal that is about four million cycles (Hz) wide. As I explained earlier, most audio tape recorders (with stationary heads) can only record up to around 18,000 Hz. To record lengthwise on the tape would require horrendous forward speed. Most audio home recorders have a fast speed of 7½ inches per second (IPS). Even studio sound recorders rarely run faster than 15 inches per second. If you consider that audio cassettes run at 1 ⅞ IPS forward speed and compare that with what's needed for fast-scan video, it's easy to understand the theory behind *slant track* and *quadruplex* recording. In order to achieve the bandwidth necessary for video recording, the forward tape speed must be very high. I've heard arguments that a tape speed of up to 14,000 inches per second might be necessary. The exact figure differs with the particular source, but if a stationary head were used for FSTV recording, the forward tape speed would result in prohibitive costs. So, a method had to be found to "write" information onto the tape in a way that would permit slower forward tape speed, higher information recording density, and thereby require less tape.

The first system to gain wide acceptance was the *perpendicular,* or

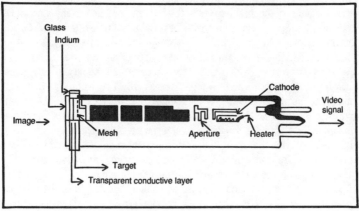

Fig. 4-5. Cutaway view of a Vidicon tube.

quadruplex, system developed by Ampex Corporation. The name quadruplex is derived from the fact that the record/playback head units consists of four individual tiny electromagnets on a single concentric shaft that is rotated in excess of 14,000 rpm. The heads are space at 90-degree intervals around the shaft or *drum* as it is known, and the tape moves at a forward speed of 15 inches per second. The same amount of information can be recorded using this system at a relatively low speed as could be recorded with a stationary head and very high tape speed. This combination of forward tape motion and head rotational speed is known as the *tape writing speed,* and it's an exacting standard.

Well, Quadruplex machines were the first big breakthrough, but they were quite large and not exactly what you might want in your living room. The Ampex Model 2000, which is still one of the industry backbones, weighs about 1300 pounds and is about 10 ft long, 6 ft high, and the depth of a living-room couch. They have their place, and that's in the studio or production/editing center. They are not cheap to run since they draw more electricity than a window air conditioner and use expensive 2-inch open-reel tape. A new one-hour roll of 2-inch tape is about $280 these days, and the price keeps going up.

Meanwhile, Sony and several other firms were working toward developing another recording format using narrower tape and an even slower forward tape motion speed. This was dubbed *slant track* and later became known as *helical scan.* The reason for these terms will become evident shortly. Slant track recording really took off in the late '60s when Sony introduced their ¾-inch U-Matic format recorder. This was an "industrial" unit that they hoped would spin off to the consumer world. Some consumer videophiles eagerly grabbed up these goodies, but so did the broadcast industry. Soon these machines were found in almost every TV station in the United States. Initially they were not used on the air, but for such things as offline editing and preview screening. With the introduction of the U-Matic recorder/player, the death throes had begun for film in television production.

Unfortunately, few people outside the broadcast industry, educational institutions, and electronics industry were willing to purchase these new machines. They were fairly large, but would easily fit atop any console television receiver. Sony, Panasonic, JVC, and others brought out consumer-oriented versions equipped with vhf/uhf tuners and automatic timers for delayed recording. The public just wouldn't bite. Soon, the U-Matic recorder manufacturers realized that their sales would be to industry, broadcast, and educational uses. New models were designed with these markets in mind, as were the peripherals (such as editing and timebase correction equipment). Over the years, skilled engineers and technicians have modified these machines for a myriad of reasons and in many ways. Manufacturers provided the basic machines, and someone else made them perform in ways that the manufacturer didn't dream possible. Eventually, the manufacturers (with Sony leading) did catch on, and such items as the portable U-Matic VCR and complete plug-in editing accessories were introduced. As time passed, the price of the U-Matic recorder went out of the average home electronics consumer's price range.

The U-Matic concept differed from the quadruplex system in a number of ways. The most obvious difference was the way the tape was packaged in a sealed container or *cassette* which the machine itself opened and threaded. In quadruplex operation, the machine is threaded manually. Also the forward tape speed was reduced from 15 inches per second to 7½ IPS—half the tape, yet the same recording time. These were the exterior differences. What about the electronic differences? There were many. Just the size difference between quadruplex and U-Matic dictated this. The Japanese, who truly deserve most of the credit for developing slant track recording, were able to engineer small circuit boards and made extensive use of integrated circuit (IC) chips thus eliminating many discrete components. Also eliminated were two of the heads. Slant track recording as developed by the Japanese electronics industry had standardized a two-head configuration for standard record/playback configuration. Some industrial/educational slant track machines did have a four-head configuration to facilitate editing. There were variations in recording standards in the early days, but with the adoption of the EIAJ-1 international agreement, tape interchangeability among machines of different manufacture became a reality. An example of this is my reel-to-reel Sony of the late 1960s. Tapes recorded on this machine can only be played back on another machine of the same type. It was manufactured before EIAJ-1 was adopted. This means that I can record and play tapes on this unit with no trouble, but someone else can't take my tape and play it on their machine unless it was built to the same specifications, when EIAJ-1 became the norm for record/playback purposes, this type of problem disappeared. By the time Betamax, U-Matic, and VHS formats came on the scene, standards had been adopted for them as well. No matter what the configuration of a machine, a tape recorded on one could be played back on any other unit of the same format. For home recording, the story really ends here but I will take it a step further in relation to the broadcaster.

Earlier in this book I said that a tape format called *Type C* was quickly replacing the venerable 2-inch Quadruplex which had been the broadcast industry standard for years. There are many reasons for this: lower tape costs, easier tape handling, lower maintenance costs, and possibly an improvement in overall picture quality. The last point is still heavily debated. What I find significant is that unlike the past when manufacturers went their own way in developing new standards, this time a worldwide agreement on standards was formalized prior to Type C conceptual design. Just like your Betamax, VHS, or U-Matic, tapes recorded on any direct color Type C NTSC standard machine can be replayed on any other machine regardless of who manufactured it. The same is true for PAL and SECAM versions of these machines—definitely a giant step forward.

Chapter 5

Some Made It, Others Didn't

In a market keen for the consumer dollar, there will always be winners and losers. Thus far, two systems have survived in the consumer electronic video recording marketplace. Were they the only entries in the "big dollar race?" No. Over the years, there have been numerous others, and it all began with open reel-to-reel black-and-white FSTV.

Even in the early 1960s you could purchase a video recorder for use in your home. You could record black-and-white programs on a ½-inch open-reel tape and play them back, if you were dedicated and content with a somewhat fuzzy representation of the original. You couldn't exchange tapes with others, nor did these machines have built-in tuners for off-air recording or integral timers for time-shift recording. While they might have been thought of as consumer oriented, only the manufacturers thought this way. The public didn't, and not much happened on the consumer video recording front until the late '60s when the U-Matic came on the scene. What did the U-Matic have going for it?

U-MATIC

First, the machines looked "smart" and "sharp." That is, they had good-looking cabinetry enclosing their inner works. The U-Matics were cassette load rather than the more cumbersome open reel design. The user only had to pop in a tape cassette and punch a few buttons. Consumer-oriented machines such as the Sony VO-1800 and JVC 6300 had built-in tuners for the vhf and uhf television bands as well as timers for time-shift recording. (In the Sony version the timer was an outboard option, while the JVC used an integral battery-operated clock.) Still it was a "no-go" for the consumer dollar with the U-Matic. Why? There does not seem to be a truly clear-cut answer, but let's try to put the pieces together. For one thing, like its open-reel predecessor, the U-Matic format was limited to a maximum of one hour record/playback time. This meant a person would have to be at home to switch tape cassettes in order to record any program longer than one hour. So much for recording that "2-hour special" while out playing eighteen holes of golf. Then there was the price of the cassettes themselves: a 60-minute U-Matic tape cassette cost about $30.

There was also the equipment availability problem. Most consumer electronics outlets did not want to make a heavy investment in an unproven (to them) item. After all, they would be the ones who would pay up-front, and they would be the ones to take a loss if the public didn't bite. Because most consumer electronics firms shied away from carrying this type of product, anyone who wanted a machine of this sort was forced to deal with a company specializing in the sale of such equipment to educational or industrial organizations. There was little public exposure of the U-Matic, and as a result many never knew it existed. As stated earlier, the broadcast industry did see a lot of potential in the U-Matic and it became an industry standard.

BETAMAX (BETA)

Sony's original Betamax did a bit better. It did have an integral vhf/uhf tuner, accessory timer, and produced exquisite "pictures", but it was still hampered with the one-hour maximum time limit. The introduction of longer playing tapes (L-750) eventually extended the record time to an hour and a half, but that was a far cry from what the American public wanted. By this time, many TV specials and feature movies were three- to four hours long, sometimes segmented on a day-to-day basis. It was not until Sony introduced their model SL-8200 Betamax that the format really took off. It offered up to three hours of record time, which meant that a complete movie would fit on a single tape. The SL-8200 was the "transition" recorder because it would play and record at the older one-hour speed (Beta I) or the newer two-hour (Beta II) speed. Sony's engineering minimized picture degradation at the slower speed. It also meant that someone stepping up to a new machine could still play back tapes recorded on an earlier version. The SL-8200 brought Betamax format recording to the forefront. With the introduction of the five-hour Beta III in late 1979, the Beta format recording system was firmly entrenched in the battle for the consumer market share.

VIDEO HOME SYSTEM (VHS)

VHS which stands for Video Home System is another success story. If fact, VHS which was introduced to the U.S. marketplace only a year or two after Betamax almost wiped-out the latter from the marketplace (if you believe what dealers and distributors tell you). With one of America's biggest consumer electronics manufacturers at the forefront, VHS hit the market with a blitzkreig of advertising that soon made the three letters "VHS" a common household word.

It's almost ironic that one of this nation's leaders in consumer electronics had turned to Japan for a home videotape format, but when the machines hit the U.S. marketplace the public gobbled them up. Why? In low-end standard configuration the machines offered a choice of two speeds with up to four hours recording, a built-in tuner and timer, and a more compact size than the competition Betamax machines. Keen advertising and merchandising made VHS the front runner in the consumer electronics race in a very short time. For a while, dealers were reporting sales ratios of 2:1, 3:1, or greater, of VHS machines over the Beta-format competition. Eventually as already stated, Betamax was re-engineered to extend recording time and added consumer-oriented bells and whistles, resulting in their regaining a good share of the market.

LOSERS

VHS and Betamax are here. Both are accepted nationally and probably will become worldwide standards as well. But what about those that failed? As I said, in a competitive marketplace, some will win and others will lose. Here are some of the casualties.

One of the best of the early cassette formats was the Sanyo V-Cord. This machine was a true gem, with the best points of the day going for it: up to two hours recording, two speeds, and an enclosed tape cassette. The machine was as easy to load and use as either VHS or Beta. The tape transport resembled a Betamax transport turned sideways and it used the same "omega tape wrap system". It had an integral vhf/uhf tuning system, but the timer for time-shift recording was an option. Why didn't it catch on? Most blame this one on dealer resistance. Nobody knows for sure. It was released about the same time as the original Betamax, but got little publicity: that probably had a lot to do with its poor market showing.

It is generally believed that the Quasar "Time Machine" failed for the same reason. An expensive item with no sure market, it kept dealers from stocking and advertising the machine. While Betamax and VHS had big-buck advertising budgets, the "Time Machine" was apparently never given the same push. There was a big initial advertising campaign, but I can remember no follow-up. The machine was good with performance equal to VHS in resolution. It also had the necessary bells and whistles of the day, yet the "Time Machine" has been all but forgotten.

Another early '70s entry was by a company named Cartravision, of San Jose, California. Their machine was a definite departure from what the Japanese were building. Unlike the Betamax or VHS machine of today which you can consider an expensive "add-on" to your present TV set, the Cartravision VCR was designed to be an integral part of a television receiver. It was a vertical-mount affair about 22 inches high and more than a foot wide. It used three heads in a system that recorded alternate lines of video and replayed each line twice. This system is called skip field recording.

The Cartravision entry was more than just a deck. It used a ½-inch cartridge with up to an hour's record/play time and incorporated a 24-hour timer. Since it was not designed to "stand alone", it depended upon the television's tuner and signal-processing electronics for off-air recording.

Unlike V-Cord and Quasar's "Time Machine", the Cartravision system got a lot of national exposure, mainly from Sears Roebuck (for about a year and a half). The unit appeared as part of a beautiful Sears top-of-the-line console TV which featured their "top chassis and picture tube." The set was even shown on the Johnny Carson *Tonight Show* coast-to-coast one Christmas season. Again, nobody can pin it down, but the Cartravision system went away rather swiftly. At the time the Cartravision unit was on the market, I was employed by Sears, and hanging on my wall is a certificate dated February 5, 1973, which attests to my training by Sears to service these machines. In the years that followed I saw very few Cartravision VCRs.

There were other entries in the race, but V-Cord, Time Machine, and Cartravision are the only three that even made a small dent compared to the "big two" in the marketplace. They all shared a common heritage, being variations on the theme of helical-scan slant-track recording. Over the years, only the Betamax and VHS systems have survived the big dollar tug of war—the rest of this volume will deal with these two systems.

Chapter 6

Betamax versus VHS

In order to discuss the differences between the two systems, we must consider several categories. Some of the items we will cover are tape writing speed, threading and tape handling, recorded track width, video head gap, and other technical aspects. I also present "subjective judgment," that is to say, "the way the picture looks to the average viewer on a properly operating consumer receiver when judged against some standard." In this category, the "standard" will be a Sony Standard Test Tape played back on a Sony Model 2800, U-Matic format machine (¾ inch) into a Sony KV-1920 table-top receiver. The U-Matic format was chosen as the standard since its wider tape width guarantees a minimum of 320 lines monochrome horizontal resolution, and 240 lines chrominance resolution. The forward speed of this type of machine is 9.53 cm per second (3¾ ips). Prior to conducting the subjective test, all machines were checked and adjusted to manufacturers' specifications usng recommended procedures and equipment. The machines used in this subjective test are my own, and they receive above average presentive maintenance. (Note: The U-Matic was borrowed for this test, but prior to use was subjected to the same technical set-up.) The results of subjective testing appear at the end of this chapter. First, the technical stuff.

BASIC BETAMAX (SONY MODEL SLO-320)

Every video recorder, regardless of format, must have certain basic operating controls for tape transport and record/play electronics. Since we want to consider a basic tape transport at this point, let's glance at Fig. 6-1. Pictured is a Sony Model SLO-320 industrial Betamax format recorder. Since these machines are designed primarily for industrial/educational use

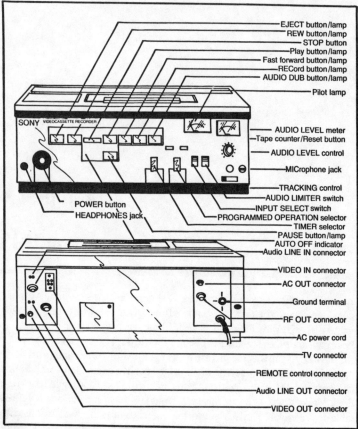

EJECT button/lamp
REW button/lamp
STOP button
Play button/lamp
Fast forward button/lamp
RECord button/lamp
AUDIO DUB button/lamp
Pilot lamp
AUDIO LEVEL meter
Tape counter/Reset button
AUDIO LEVEL control
MICrophone jack
TRACKING control
AUDIO LIMITER switch
INPUT SELECT switch
PROGRAMMED OPERATION selector
TIMER selector
PAUSE button/lamp
AUTO OFF indicator
Audio LINE IN connector
VIDEO IN connector
AC OUT connector
Ground terminal
RF OUT connector
AC power cord
TV connector
REMOTE control connector
Audio LINE OUT connector
VIDEO OUT connector

POWER button
HEADPHONES jack

SONY VIDEOCASSETTE RECORDER

Fig. 6-1. Sony Model SLO-320 Beta-format recorder.

with a video camera, the usual bells and whistles found in a home machine have been deleted. Don't be fooled by the seeming lack of goodies. Industrial-grade machines are far more expensive than normal home machines, since they must be built to withstand more traumatic conditions than they would suffer in the consumer market. The SLO-320 was chosen for this discussion because it does represent a basic control configuration without the goodies found on even the least expensive home machine. The real differences are the addition of an audio level meter and an audio gain control, not found on consumer machines.

Looking at the front panel layout, left to right, we first find the power on-off switch. Adjoining it is an earphone jack for monitoring audio being recorded, and audio playback where private listening is preferred. It's also very useful in tape editing.

Next comes the "transport operation center." That's the row of controls (buttons in this case) in the center of the panel. To the very left is the

"Eject" button. Depending on the design of your machine (control sophisti-cation) it can be used to remove the cassette from the machine once the recording cycle has been completed, to initially load the machine, or to remove a partially recorded cassette.

WARNING: On some machines that use a mechanically activated trans-port, pushing the "Eject" button while the tape is in motion (Record, Play, Fast forward, or Rewind) can cause severe damage to the recorder/player. Most recorders using mechanical control have designed-in lockouts to prevent acci-dental use of this function at the wrong time. Other machines with electro-mechanical sequencing permit the "Eject" control to be utilized in all modes other than record. If you are not sure which category your machine falls into, either consult your owner's manual or place the unit into "stop" prior to ejecting a cassette. This simple precaution can save a lot of expense for unwarranted repairs.

The next control is the "Rewind" button, and as its name implies, its function is to activate the rewind mode once the tape in the cassette has been exhausted. The "Stop" button does just that: It stops the action of the tape transport and disables part of the recorder's electronics by removing power from certain circuits. In a VHS unit, it will also cause the tape to be unthreaded and returned to its cassette. Betamax format machines keep the tape threaded in the transport. This difference will be discussed in more depth later.

Pushing the button marked "Play" will activate the tape transport and the playback electronics, providing the signal recovered from the tape to the output jack(s) on the rear panel. The "Fast Forward" button is the opposite of the "Rewind" button. Pushing it activates a high-speed forward tape movement condition for locating a specific point on the tape for playback or record.

Depending on the design of the machine, the record function may be activated in one of three ways. Some machines only require pushing the "Record" button to activate this mode. Others, specifically older VHS and Betamax units (SL-7200 by Sony and VBT-200 by RCA) required that you simultaneously press the "Record" and "Play" buttons to activate record mode. Still others have a separate "Record Lock-Out" control that you must engage before pushing "Record." The last two techniques are to preclude accidental erasure of recorded material. They cause you to stop and think a moment before hitting the "Record" button. With recorders, audio or video, there is no retrieving material once it has been erased. Other safety devices to prevent accidental erasure are built into the tape cassettes and will be discussed later.

Except in the very "low end" of consumer machines, most home recorders now include an "Audio Dub" button. Its purpose is to activate the audio erase/record electronics while not affecting the recorded video material. This gives the user the option of removing an unwanted audio track and replacing it with a new one. For example, suppose you have transferred some of your home movies to videotape. Many people are doing this these days, usually by sending the film to a specialty house that performs this task.

It's no secret that most home movies are either 8mm or Super-8mm film and have no sound. You have made the move to videotape and after viewing it a few times you realize something is missing—sound. The "Audio Dub" control permits you to add music, commentary, or whatever you might want to give more realism to your videography. As another example, you may have a tape with an audio track you don't like. Again, the "Audio Dub" button will let you change things to suit your taste.

In the case of the SLO-320, the "Pause" control is located just below the "Play" button. Its position will vary from machine to machine and from manufacturer to manufacturer. Its purpose is to stop the motion of the tape in either record or play modes without disengaging either the tape transport or the electronics. To the average consumer, the button has two important functions: First, it can be used to edit out commercial breaks in a program being recorded off-air. Second, when used with a video camera, it permits on-site editing of each scene. It's bad practice to leave any video recorder in "Pause" mode for extended periods. This is because the tape motion has stopped, but the scanner head is still passing over the tape at its normal speed. Extensive operation in this mode can lead to tape damage, clogged heads, and shortened scanner assembly life. A good rule of thumb in using the "Pause" mode is to consider 7 to 10 minutes as an absolute maximum. Some newer machines have incorporated a timing circuit that will automatically return the machine to "Stop" if the user leaves it in the "Pause" mode for too long.

To the right of the transport controls you will see a digital counter. This device is included on most VCRs to help locate an approximate point on the tape, either for playback or to insert new material. Note that I said "approximate" rather than "exact" point. This is because tape index counters are usually mechanically driven by the supply or take-up mechanism, and their accuracy is marginal at best. They will get you to the "ball park," but the only accurate way to find an exact point is to advance the tape using the counter to the approximate position, and then play back and view until the exact spot you want is located. Some of the newest machines coming to the consumer market are changing to an all-electronic tape index counter. This one counts control-track pulses recorded on the tape as part of the normal recording process. There is a third, highly accurate method of tape indexing, but it is not yet available for consumer use. The system, called "Time Code," is a digital signal containing hour, minute, second, and frame information that is recorded on an audio track. This system is used in the broadcast industry to permit "field frame accurate editing." It requires that the video recorder have at least two audio tracks available. Multiple audio tracks are standard on U-Matic, quadruplex and 1-inch helical broadcast machines. They are also available on some industrial recorders of the

Betamax or VHS designs. Multiple audio tracks seem "down the road" yet for home video, and when the system is incorporated, it will probably be used for either stereo sound or multilingual presentation. There is another limiting factor: cost. Time Code generators and readers are quite expensive, from $4,000 to $10,000 or more in "add-on" equipment. In broadcasting it's a necessary tool. For the consumer, it would be a rather expensive luxury. So much for the basic video recorder front panel. (The rest of the controls on the Sony SLO-320 front panel are specialized and we will ignore them for now.)

A quick glance at the rear panel (Fig. 6-1) will give you an idea of what you will find in a basic video recorder/player—no bells and whistles. On the left side of the panel (top to bottom) is a row of connectors. Since this unit has no internal tuner, these jacks are used to bring signals from a camera or other video/audio source to the machine and to take output signals from the machine. These signals are usually at standard levels: video at 1.0 v peak-to-peak across a 70 ohm unbalanced line; audio at −10dB per 100,000 ohms input and −5dB across 100,000 ohms output (both unbalanced).

The next vertical row of jacks are specific to the SLO-320 and will not be discussed (they are used mainly for production and editing). Skipping across to the righthand side you see a type **F** connector marked "RF Out." This permits the recovered signal to be replayed on any TV set, simply by running a cable from that point to the antenna terminals on the TV set and tuning to channel 3 (or 4 depending on the machine's rf modulator setting).

This machine also has an accessory socket marked "AC Out." This is to connect other devices that require 110 Vac power. Do not exceed the recorder manufacturer's specified power ratings. Finally, you will see a "Ground Terminal." Some machines have them, others don't. Consult the user's manual to determine if your machine requires an external ground connection. Except for an ac power cord, that's it for describing the basic recorder/player.

To use a VCR such as this for recording requires connection to an external source of video and audio. This might be a camera and microphone, another video player, or an external tuner unit.

BASIC VHS (GENERAL ELECTRIC MODEL 1VCR2002X)

The General Electric Model 1VCR2002X (Fig. 6-2) is a basic home VCR of the '80s. The transport control configuration has been moved around a bit (compared to the Sony SLO-320), the "Record, Rewind, Play, Fast Forward, Stop, and Pause" buttons are still neatly grouped together. The "Eject" button is located a bit lower on the machine's front panel. The on-off power switch is in the lower left corner. Slightly below and to the right are a number of other jacks and controls. Let's look at them, left to right. First there is a micro-jack marked "Pause." This permits connecting a remote-control pause button that can be operated from the viewing position. The video and audio input/output jacks that were on the rear of the Sony SLO-320 are conveniently front mounted on the GE machine. To the right of them is a switch marked "TV/VCR"—to select the signal sent to your TV set, from either the antenna or the video tape.

For this discussion, I will assume that the signal source for both your TV set and VCR is an antenna on the roof of your house (see Fig. 6-3). Cable television interconnects will be discussed later. Most modern antenna installations use a single, split-band antenna designed to receive vhf low-band (Channels 2-6), vhf high-band (Channels 7-13) and uhf-band (Channel 14-83) signals and transmit them to the TV or VCR on a single download, or *transmission line*. Today, shielded 75 ohm coaxial cable is replacing the older 300 ohm twinlead used for many years. Coaxial cable is less suscepti-ble to noise and unwanted signal pickup, and it handles environmental changes better than most forms of twinlead.

Assuming the use of an "all-channel" antenna with 75 ohm coaxial download, both vhf and uhf signals are present on the same line. They must be separated by a *band splitter* that provides two outputs. One output is vhf (high and low bands), and the other output is the uhf band. These two outputs are connected to the proper antenna input terminals on the VCR. The vhf and uhf signals are routed inside the VCR to their respective tuners and to a separate set of contacts on the "TV/VCR" switch. With the switch in the "TV" position, the antenna's vhf and uhf signals are routed to the appropriate output connectors on the VCR rear panel and then to the television set antenna terminals. In this switch position, you can view one TV channel normally and record another channel for later viewing, because the antenna signals are always connected to the VCR tuners.

Suppose you want to either view something being recorded or play back a tape. The signal source now becomes the VCR itself. Flipping the switch to the "VCR" position connects the vhf-out line of the VCR's internal rf modulator to the vhf output terminal. The rf modulator takes the video

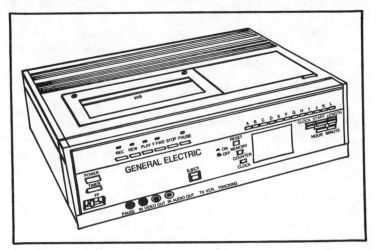

Fig. 6-2. General Electric VHS-format recorder.

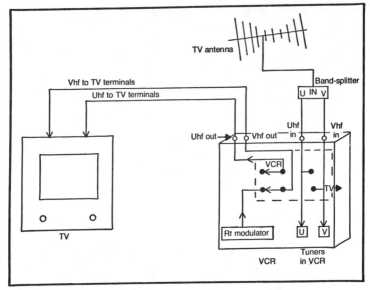

Fig. 6-3. Typical VCR installation.

and audio signals from the VCR and makes them into a TV signal again, sort of like a micro-power TV station transmitter inside your machine. You now set your TV to the channel prescribed in the VCR owner's manual. If you are in playback mode, you will see recorded material from the tape itself. In record mode you would see the electronic-to-electronic monitoring the signals going to the video and audio record heads. This is called "E-to-E Monitoring." You would see and hear the signal **as it is being recorded**, not as it looks on the tape. Only the multi-multi-kilobuck broadcast machines have "off-tape" monitoring while recording. This is done by installing a separate "confidence playback" head on the scanner assembly and routing its signals through a separate set of amplifiers to a monitor. No home VCR has this ability—at least not yet. The "TV/VCR" switch however, makes it possible to record one station while watching another, without a second antenna! It is a simple but effective device.

The last control in the GE VCR's "lower compartment" is a knob marked "Tracking." Its precise function will be discussed later, but in essence it permits the VCR user to optimize the machine's operation to produce the best quality picture.

The black square to the right side of the panel is an internal digital clock which is used to pre-program the time shift for delayed recording. It also serves as a tape counter on this particular VCR model. The control buttons to the left of the clock control the counter and counter memory, while those to its right set the clock and control the time-delay recording cycle. The row of buttons above the clock contains the station selector buttons. This unit uses all-electronic varactor tuning (rather than conventional mechanical tuners), and simply touching one of the buttons changes the station being received. Most current VCRs have gone to this type of vhf/uhf tuning arrangement since it improves reliability by eliminating mechanical parts. In more sophisticated microprocessor-controlled units, it also makes for easier circuit design compatibility.

While not shown, the rear panel of this VCR has two jacks for antenna input and two for antenna output—one set for vhf and one set for uhf signals. (Note the output impedance on the GE Machine is 75 ohms, but external conversion transformers for impedance conversion are provided). GE includes a 16-foot remote pause control cord and a blank cassette with this package, so you can just take it home, hook it up, and start using it immediately. So much for outward appearance, now onto the tape transports themselves.

TAPE HANDLING

The big difference between VHS and Beta-format machines is in how they load and handle tape in the transport. As I stated earlier, both systems are variations on Sony's original U-Matic theme, but neither handles tape in exactly the same way as a U-Matic format recorder. To start with, in U-Matic, the supply reel is on the right with takeup reel on the left. Tape motion is basically counterclockwise. Both VHS and Betamax VCRs supply tape from the lefthand reel, and the takeup reel is on the right—basically a clockwise movement. (In none of these cases does the tape travel in a full circle back to its initiating point.) All three systems use cassette-encased reel-to-reel tape cartridge designs, as opposed to something like eight-track audio which is an endless-loop single-reel system.

In a Betamax, (Fig. 6-4) threatening takes place when the cassette is inserted into the machine and the cassette holder is depressed. Once threaded, the tape is never retracted into the cassette until the offload cycle is initiated by depressing the "Eject" control. Tape motion is always through the transport mechanism even in fast forward and rewind shuttle modes. So that the machine will know when it has reached the limit of tape travel for a given cassette, leader and trailer tape in the cassette is coated with a metallic substance. A detector in the machine is coupled to a set of sensors that detect the presence of the metallized tape and automatically stop tape motion. In the Betamax system, these sensors are like small radio antennas that are sharply tuned to a specific radio frequency. When the metallized tape passes near either sensor, they are detuned and upset certain circuits in a predetermined way. This disturbance generates a logic condition that says it's time to stop the tapes.

Note that the tape forms the letter "U" or, as some say, an Omega sign around the scanner. The tape is extracted from the cassette by a manipulator on the machine's loading ring and wound 270 degrees around the scanner unit. Here's where things get a bit complicated though. While the tape motion forward remains clockwise, the tape is loaded around the scanner in a counterclockwise direction. It takes about 25.5 inches of tape to completely load a Betamax format machine. Let's follow the tape path as shown in Fig. 6-4.

The tape exits the cassette on the lefthand side and is drawn into the machine. It first passes a series of tape guides and the full-track erase head before wrapping around the scanner. As it begins its trip around the scanner, the tape's angle of travel with regard to a horizontal plane changes. It continues this angular change around the scanner and past the audio/control track heads (carefully aligned in azimuth to compensate for the angular tape motion) until it reaches the capstan/pinch roller. Here, it levels to the horizontal again and with the help of more tape guides performs a 180-degree change in direction and then another 90-degree change for a return to the takeup reel in the Betamax cassette. Like it's U-Matic brethren, the Betamax transport, although designed with a minimum of moving parts, is a complex piece of electromechanical machinery.

Figure 6-5 shows the basic VHS tape handling system. It too is an offshoot of U-Matic although it differs markedly from the "U" pattern of tape travel. In fact, it forms the letter "M" and has become known as *M-wrap*. The basic M-wrap concept is a Sony design, and if you look at the early open-reel half-track machines, you will realize that some were an inverted **M-wrap** pattern. More like a "W" with a downward lump in the middle.

In the M-wrap system, tape is pulled from the cassette, passes some

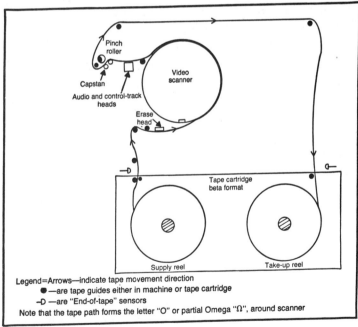

Fig. 6-4. Beta tape thread configuration.

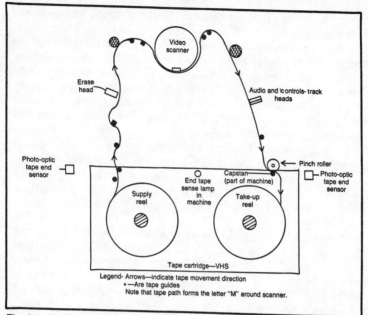

Fig. 6-5. VHS tape threading configuration.

tape guides, and passes the erase head. So far there's little difference from Betamax, but this is where the similarity ends. The tape now travels to the scanner by way of angular alignment pins which set the proper angle for the tape in relation to the scanner. The tape is wound around the scanner for 205 degrees and exits past another angular alignment pin which resets the tape path to the horizontal travel plane. It then passes the audio- and control-track heads, passes the pinch roller/capstan combination, and exits to the takeup reel of the VHS cassette. It takes about 13.3 inches of tape to load a VHS machine, and the tape does not remain threaded in the machine while in standby modes (except pause).

Tape threading takes place only during record and play cycles. During shuttle modes (fast forward and rewind) the tape remains in the cassette and the shuttle process takes place through that shortened route.

Which is better for tape? After speaking with a number of tape suppliers and technicians, I conclude that both systems are about equal, and that today's tape quality is such that loading stress is not really a concern anymore. If machines of either type are properly maintained, neither system should ever give much of a problem.

The end-of-tape sensing system in VHS is actually a bit simpler than that used in Betamax units. The sensors are photo-optic cells that are activated by a light source provided by the machine (similar to U-Matic format machines). The VHS cassette has precision holes molded into it that align with the activator lamp and sensor cells in a horizontal plane (see Fig. 6-5). The leader tape on a VHS cassette is clear, and when it passes between the light source and sensor cell it permits an optic path to be completed. When either sensor sees light, it does exactly what its Betamax counterpart would do. It generates a logic output that says to the machine, "Stop."

Which end-of-tape sensor system is better? I can only say that as a service technician the only problems I have really found with the VHS system were the lamps themselves.

Eventually any incandescent lamp will burn out. Even that's becoming a problem of the past because newer VHS designs are using solidstate lamps called light emitting diodes (LED) as a light source. Unless the LED is subjected to excessive physical or electrical abuse, its life expectancy is probably longer than those who made them.

After years in the videotape game, and speaking objectively, I cannot find fault with either method. Both systems work amazingly well. Also, if you have a VHS machine with an incandescent sensor lamp, don't worry if it goes bad. You won't find yourself breaking expensive cassettes, since the design of the control logic precludes operation if the lamp dies.

Here's a quick service note to VHS owners: If your machine seems to power-up but won't operate, look to see if the lamp in the cassette compartment is lit. On some portables you have to insert a cassette to do this, since the lamp is pulsed on and off to conserve battery life. If the lamp refuses to light at all, chances are it's burned out and that's why the recorder just sits there like a $1000 bump on a log. A word of caution with this, however: If the bulb is bad don't try to replace it yourself. Let a qualified service technician do it. It's not as easy as it looks. Take it from someone who knows!

Chapter 7

The Video Head

If a video recorder has a "heart" then a component known as the video head assembly, or *scanner*, is just that. One might say it's the point where "tape meets machine" for the video portion of the record/playback process. At first glance the scanner looks simple enough, but don't be deceived. It's a precision component built to rigorous specifications and cannot tolerate mistreatment. Whether you are talking about Betamax, VHS, U-Matic, or any other helical format, the scanner must be treated with kid gloves. Even a slight scratch on its surface will render the scanner useless.

Staying with the basic two-field helical record/play process, the element containing the recording heads (upper drum assembly on VHS or scanner disc on Betamax) rotates at a precise speed, driven either by the recorder's main motor or a spearate scanner motor. The rotational speed is precisely 1,800 rpm. (This speed and head position are critical in playback, as will be explained later.) As each of the two heads passes diagonally across the surface of the tape, it either records or recovers a theoretical 262½ lines of video information. Each time the scanner rotates a full 360 degrees, a complete 525-line frame of information transfer takes place. This holds true for both systems, but there are differences which do effect the perceived quality of the reproduced information: two key terms are *head gap* and *track width*.

For example, my first Betamax format recorder was the Sony SL-7200. I still have it. In that and present-day industrial Betamax format units the gap was set at 60 microns. When Betamax II Zero Guardband format was introduced, the gap of the heads was reduced—"cut in half" to 30 microns. The advent of the 5-hour Betamax III format brought even a smaller head gap. It worked out at 19.4 to 20 microns, dependent upon machine manufacturer. A similar evolutionary process took place in the VHS camp as forward tape speed was reduced to provide longer play on a given tape cassette. The early VHS "SP" machines used a head gap between 56 and 58 microns. It then went to 45 microns with the advent of "SP-LP" machines and finally to 29.98 microns (effectively 30 microns) when the 6-hour "SLP" format was introduced. To digress a moment, in VHS the head technology has evolved even further. Some top-of-the-line machines now utilize two sets of heads on their scanner assembly. One set for "SP" record/playback (about 59 microns) and another pair for the extended-play modes. The latter are smaller: approximately 20 microns. This was done because the top-of-the-line machines feature certain special effects that perform better at differing tape speeds with different-size heads. Note however that in a four-head machine, at any given time, only two of the four heads are in use. Selection of the proper head pair takes place manually in record mode when you select the speed, and automatically in playback mode since modern machines are designed to sense the proper playback speed directly from the recorded tape. Machines with special effects features will be discussed later.

One of the reasons for the larger head gap in early single-speed machines had to do with overall track width. A track is one physical line of information recorded on tape that contains 262½ lines of electronic information. The original Betamax I format utilized a fairly high forward tape speed—precisely 4 cm per second. (As a point of comparison, forward tape speed on a U-format recorder is 7½ inches per second.) Head **A** would record a line, the tape would move, and then head **B** would record its line. Tape forward momentum created a "no man's land" between the two lines of information. This area was called a *guard band*. Its purpose was to ensure that information in Line **A** would not interfere with information in Line **B**. The system was effective but also wasted a lot of tape that could be utilized for picture information if some other method could be found to fool the recorder into believing that a guard band was there even if none existed. Enter "Azimuth" recording (see Figs. 7-1 and 7-2).

As can be seen in Figs. 7-1 and 7-2, the difference between guard-band and *zero-guard-band* recording is obvious. In the latter, track **A** literally touches track **B**, which touches the next track **A**, and so on. If a zero-guard-band tape were replayed on a machine not equipped to handle it, there would be crosstalk from track to track with resultant noise in the video and color. The picture would look like garbage.

Why not just make the heads smaller? Two reasons: First there are physical limitations on "head gap", and secondly, even a smaller head is governed by certain laws of physics. Crosstalk would be minimized but still evident. Again it was Sony that solved the problem—not only for its own Betamax format, but for VHS as well. The system is called *Azimuth helical recording* and through cross-licensing, all manufacturers have reaped benefits from the concept.

Here's how the system works (see Fig. 7-3). As most audiophiles

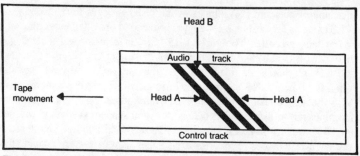

Fig. 7-1. Early Beta I-format recording with guard bands between video tracks. Shaded areas indicate video tracks. Unshaded areas are guard bands between tracks.

know, there are two important considerations in audio tape playback other than electronics. First is *head height* to ensure that the head sees the proper track on the tape. The second consideration is *head azimuth* which ensures that the entire head surface will contact the tape track to produce all frequencies recorded thereon. The azimuth concept is the basis for zero-guard-band video recording.

Suppose we cant each head a few degrees from the vertical axis, in different directions. We are creating an azimuth error on purpose, but instead of being a liability, it becomes a tool. In a video recorder, the track adjacent to a track being replayed at a given time is always a track that was recorded by the opposite head (the one not in use at the time). The two heads are separated by 180 degrees of head rotation. If the two video heads are designed with a built in azimuth error that's equal and opposite, then crosstalk can be eliminated. This is because head B will ignore high frequency overshoots from the adjacent track A. It just does not see them. The azimuth error utilized in Betamax format is a precise figure. It's $-7°$

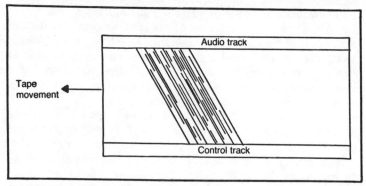

Fig. 7-2. "Zero guard band" recording used in Beta II and III and VHS systems. No blank spaces are left between video tracks. Guard bands are created mechanically and electronically in the machine.

38

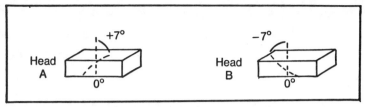

Fig. 7-3. Azimuth displacement of record/play video heads.

for the **B** head and +7° on the **A** head, for a total of 14° introduced azimuth error. That's only part of the story.

In playback, signals derived from track B are amplified as recovered, while those from track A are inverted in phase. By doing this, still more error is introduced electronically to prevent crosstalk of adjacent color information. It's the combination of these two procedures that "fool" the machine into believing that a guard band exists when none really does. (Admittedly this is an over-simplified statement, but on a minimal technical level it must suffice.) Using this system, no guard band is necessary, and tape that used to go for naught is now utilized to record and replay intelligence. Azimuth recording is common to both Betamax and VHS recording formats. The figures used above are for Betamax II, but only the "numbers" are different in other formats.

Track width is another important factor. Early home VCRs used a fairly wide track width, but with the introduction of the Betamax II format the track narrowed to approximately 29 microns. Similar reductions took place in VHS as well, thus permitting even more information to be recorded on a given length of tape. It has been a combination of narrower track width, slower recording speed and azimuth recording that have helped the home VCR, regardless of format, mature to the machine we have today.

Chapter 8

The Complete Video Cassette Recorder (VCR) System

A complete home video cassette recording system includes a record/ playback deck, its associated electronics package, a tuner or other signal source, a machine control system (either mechanical or electromechanical), and a time-delay recording function block.

The block diagram shown in Fig. 8-1 is an oversimplification of the basic home recorder, but it is the way that units for the consumer are packaged.

RECORD/PLAYBACK DECK

The tape transport consists of all mechanical equipment necessary to thread, move, and rewind the tape. On it are mounted such components as the load mechanism, scanner, audio and control track heads, erase heads, plus much more. The electronics package for record and play is common to both modes of operation. That is, the same circuits are used to retrieve information from the tape that are utilized in the recording process itself. This package is usually divided into three subsections: luminence (Y) processing, color (chroma) processing and audio processing. Note that I use the term "processing." This is because, in the case of both luminence and color, the signals are truly converted, or processed, prior to recording onto tape or in the information retrieval from tape.

Directly connected to this electronics package is a subassembly called the signal source. This is either a pair of vhf and uhf tuners (with associated circuits to produce discrete audio and video signals), or (as in the case of portable units) the source can be a combination of a camera and microphone or auxiliary tuner unit. Specifics depend upon the design concept of a particular unit.

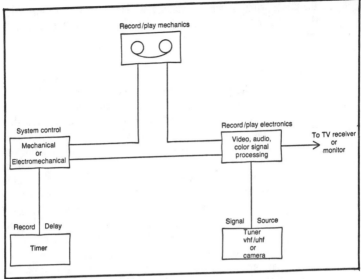

Fig. 8-1. Basic home recording system.

The timer, usually digital electronic, is used to delay recording until a prescribed time and then to end that function at another predetermined time. The timer is like a lamp timer used to turn lights on and off when you're away from home, but far more specialized and highly accurate. This unit performs its function by controlling the "brain" of the electronics—the *System Control* unit.

SYSTEM CONTROL UNIT

System control is exactly that: a unit that takes commands from an external source (the owner of the unit) and converts these commands into electronic and mechanical functions.

System control is the brain. This is because certain operations in a VCR must be accomplished in a specific sequence within an allotted time frame. Here are some of the jobs that the system control unit performs.

1. Initiation of function (thread tape, record, play, rewind, fast forward, and pause)
2. Correct sequencing of function (e.g., in loading tape, the system must assure that the rest of the machine follows the loading cycle at the proper rate and that the tape is correctly loaded)
3. Generation of control pulses in record (pulses that are recorded on the tape's control track and used for sychronizing the picture in playback)
4. Switching between modes(enable electronics and mechanics for record and play)

5. Playback synchronization (making sure that the picture locks on the screen and is free of machine-generated error) Plus . . . much much more.

In early machines, much of the system control was "mechanical electro." That is, a portion of the command was initated mechanically which actuated electronic system control circuits that would complete a given cycle and monitor the action taking place. Later machines turned that around. They are "electro-mechanical," many utilizing computer technology with built-in microprocessors that convert a single "push of a button" into a timed sequence of commands that in turn activate the different functions. It's rare to see "mechanical-electro" control any longer except in the low-end stripped-down basic machines. Take one step up from the bottom and you enter into the world of the computerized home.

For example, suppose you want to record a program. You put a tape in the unit and push the record button. In a modern unit, what you are doing is telling the microprocessor to initiate the record sequence. The microprocessor will then output a series of commands to different parts of the VCR to initiate this operation. First, if it's a VHS machine, it will tell the load mechanism to thread the tape and will control the loading cycle, monitoring to be sure the tape threads correctly and is at the proper tension. It will also make sure that the scanner is rotating, and that the tape is moving in the proper direction and at the proper speed. Once the tape is in motion it will enable the record electronics and permit signal processing to begin. Additionally it will initiate the generation of control track pulses that are recorded on the tape to permit synchronized playback. Then it continues to monitor tape movement to insure that tension remains proper throughout the recording process. If a sensor tells the system control unit that tape tension is too high or low, it will take the machine out of record and remand it to stop. In fancier machines with specialized features (such as slow motion, high-speed search, or freeze frame), it will govern these functions as well. Figure 8-2 illustrates the above.

There is a lot more to it, other than just one computer chip. The microprocessor controller is the command center: it takes your orders and decides what has to be done to comply with them. It then orders other components into operation to fulfill your desire. The nice thing about the microprocessor controller is that it eliminates many discrete parts and therefore means more rugged design.

Many of the circuits in the VCR function in dual capacities. That is they serve signal processing purposes in either record or playback mode. The system control unit is the only part, except for the power supply, that functions in all modes of operation, including something as simple as turning the entire unit on and off. On many machines, this simple action is a command to the system control unit and not a direct disconnect from the power source (see Fig. 8-3). Here, the unit's power supply is always on as

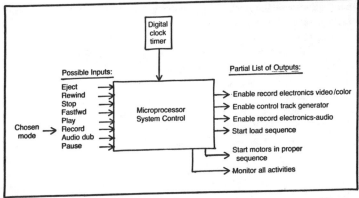

Fig. 8-2. In a microprocessor-controlled recorder, touching the "Record" button initiates all of these functions. The clock timer can do it for you if you are away.

long as the plug is in the wall. Both the power supply and system control microprocessor are activated at all times. When you turn the unit on, you are actually telling the microprocessor to connect power to the remainder of the machine. Why go to such lengths to turn something on and off? simply because in many recorders the microprocessor controller performs other functions including that of being the delayed-record timer itself. This and other circuits require constant though minimal standby power to operate when needed. Hence, electronic control has replaced the simple on-off switch found on most early VCRs.

SIGNAL-PROCESSING PACKAGE

Let's move onto the signal-processing package. That's the electronics used in the VCR to record and playback both video and audio. To discuss this aspect of VCR operation, we must first look into the basics of something called *heterodyne video recording.* This is the electronic system used in home VCRs to record and retrieve visual material. Audio program material is recorded and retrieved in the same manner as any conventional sound recording unit, although the audio material does occupy a specific portion of the tape. To explain the heterodyne recording technique again means that I must get a bit technical. I'll try to minimize the "numbers" and stay with simplified text where I can. Any numbers quoted are specifically for Betamax II format, although the principle is common to all home (and most industrial) VCRs.

In record, video from source (e.g., the VCR's tuner assembly, another VCR, or a camera) is amplified and fed through a color-trap circuit. This is done to remove the 3.58 megahertz color (chrominance) signal from the video itself. In effect, one of the first jobs the VCR does in record is to separate the black and white (Y signal) from the color information (chroma). From here, the black-and-white video signal is converted from it's

43

amplitude-modulated state to a frequency-modulated (FM) configuration since recording in this system is FM.

Let's look at the color signal for a moment. It has been separated from the Y signal video, and it then passes through a filtering network to make sure that only color information remains. After passing through more amplifiers and other specialized circuits for automatic chrominance level and phase control, it reaches a "black box" called a *heterodyne converter* (see Fig. 8-4). Here, the 3.58 MHz color is mixed with another reference signal generated in the VCR. In any heterodyne mixing system, you wind up with two possible output signals. One is higher in frequency than the original and one is lower—the sum and difference of the signals fed in. The signal used in VCRs is the lower of the two, or difference, which is 688 kilohertz (kHz). From here both the FM video and converted color signals are added together, amplified, and sent to the record heads. (Figure 8-5 is a simplified flowchart of signal-processing paths.)

In playback, the output signals from the two video heads are fed to individual amplifiers, then with the aid of an electronic switching circuit the amplifier outputs are alternately fed to a circuit called a *dropout compensator* (DOC). This is needed to make sure that information losses from the record/play process do not appear as bothersome black streaks in the final picture. The DOC stores a line of video information, and in the event that part (or all) of the next line is missing, the machine automatically inserts the previously stored line in its place. The signal, which is still FM, is then demodulated to produce an AM black-and-white picture.

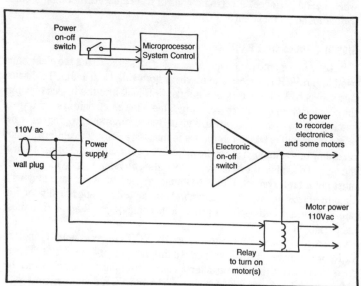

Fig.8-3.Simplified electronic on-off power control on modern VCR.Note: Power supply is always on while unit is connected to power mains.

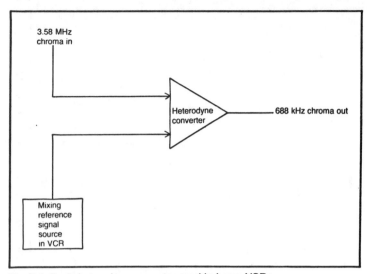

Fig. 8-4. Basic heterodyne converter used in home VCRs.

The heads also pick up the 688 kHz color signal. It is again heterodyned, but this time from 688 kHz back to 3.58 MHz, and then added to the video to produce an NTSC-standard color picture. (Note: PAL, SECAM, and Tri-Standard VCRs use different frequencies, or combinations of same, as specified by the machine manufacturer. The numbers used here are derived from the Sanyo series of Betamax II-format NTSC-standard machines.)

To be sure that the picture will lock-in and hold steady, *servo* circuits are used. Since a servo is really a circuit that compares one signal with another, it has to have a reference signal as one of its sources. It uses the

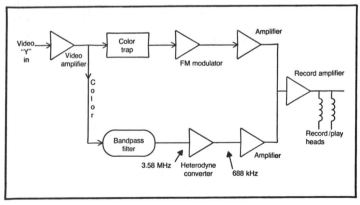

Fig. 8-5. Simplified block diagram of home VCR in "Record" configuration for video and color. Servo- and head-switching circuits not shown.

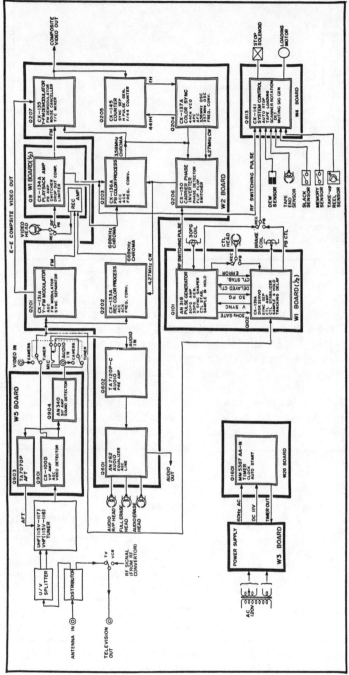

Fig. 8-6. A complete home VCR flowchart. (Sanyo Beta II System VCT-9100, courtesy Sanyo Electric Co., Consumer Products Div.).

vertical sync signal that is derived and separated from the video. The other signal is called a *30 PG pulse,* and in record it is generated electronically. In playback it comes from magnets and sensors that form a mini-electric generator from scanner rotation. The servos control the exact speed and position of the two record/play heads on the scanner so that the proper head is in contact with the tape at the proper time—for the proper amount of time. Figure 8-6 is an electronic flowchart of a typical Beta II VCR, in this case a Sanyo. By following the arrows, you can track the different circuits as they perform various functions. This chart is a very good tool for understanding the overall record/play system. The servo control is very important in the play mode so that the recorder is sure to have the proper video head in contact with the proper video track on the tape at the right instant in time. Without servo control, there would be no way to get the VCR to reproduce a steady picture (free of pulling and rolling).

AUDIO TREATMENT

Perhaps the most neglected area of video recording is that of the audio track. Here again the VCR is far better than the TV set used to reproduce the recovered material. A good number of late-model VCRs can reproduce hi-fi quality audio. The trouble is twofold: First, the sound transmitted by TV stations is far from hi-fi quality, although broadcasters are beginning to take steps toward improving this especially for music programs. Many production facilities are going toward the use of Dolby and DBX audio-enhancement systems among others in order to improve overall audio response.

On the other end of the quality spectrum is the consumer-grade TV receiver. Most portables have a single, small speaker—the same kind often found in table radios. They can reproduce sound, but the overall quality leaves something to be desired. Even many console TVs have only one or two small speakers. There are exceptions to every rule: Recently, Sanyo came out with a 19-inch portable that features 10 watts undistorted audio and a two-way hi-fi speaker system. It has loudness, bass, and treble controls just like most hi-fi amplifiers, and the audio quality is outstanding. Mitsubishi also has a 19-inch set with outstanding audio quality. Within the next few years, I suspect most manufacturers will come forth with similar designs since audio is really that last big hurdle left in the design of today's consumer video. I should note that Magnavox has always been proud of delivering exceptional quality audio on it's "stereo TV theatre" console sets. In fact, even the oldest of these is essentially ready for hi-fi TV audio.

To really enjoy the full capabilities of the audio end of a home VCR, it must be connected to an external audio amplifier. Most machines have a rear panel "Audio Out" jack that can be fed to the "Aux Input" of a good hi-fi or stereo amplifier. You won't believe the difference, especially when the VCR is run at its higher speeds (Betamax I or III, or VHS SP). This is because better high frequency response comes with faster forward tape movement. You might remember that 7¼ ips always sounded better than

3¾ ips in the days of open-reel home audio recorders. The analogy holds true here.

Also, the audio recording system is that of linear-direct record. About the only "processing" that takes place in a home VCR is automatic volume leveling and automatic selection of proper record bias for the speed chosen. The audio system in a home VCR is no more complex than the comparable home audio recorder—in some ways even simpler.

What about stereo with home video? Currently that's only available with pre-recorded material on U-Matic format tape. It is possible to include two audio tracks on ½-inch tape. The Sony industrial recorder discussed earlier has this feature, not necessarily for stereo sound. In this case, it's more likely to be used for insertion of time code for editing purposes. Also, I've yet to see any mass-marketing of pre-recorded software (films, shows, etc.) on Betamax I format. I predict that stereo will come to telecasting itself as well as home video recording, but it is way down the road yet.

At this point I will end this discussion of the home video recording technique. For those of you who are fascinated by the dynamics of the electromechanical systems used, and who want to learn more, I suggest that you look into evening training classes available at many colleges and trade schools. To go any deeper on a technical level requires a good understanding of basic electronics solid state circuitry. Both of these are far beyond the scope of this work.

Right now, it's time to look at the "subjective testing" that I have done. The next chapter will deal with my personal views of both Betamax and VHS home systems. they are based solely on personal observation and are not meant to be the final word on the subject. In the end, you must judge each system for yourself and make your own decision on what best suits your needs. Also, remember that technology keeps moving. Improvements in picture quality are coming fast. Use the following chapter as a guide, but make up your own mind when you go out to purchase a home VCR. Do this by knowing how to perform your own subjective testing.

Chapter 9

How to Buy a VCR

A few years ago, buying a home VCR was easy. You simply walked into the local TV shop and told the salesman you wanted a Betamax. Chances were he didn't even have one in stock. You plunked down a deposit, and a week or two later you picked up your new toy. It was easy in the mid 70s. With the advent of a second format, VHS, and refinements to both Betamax and VHS, it's not as easy now. Today there are probably over a hundred different machines to choose from in both formats. There are bells and whistles galore. To complicate things even further, some retailers seem to have no concept of what they are trying to get you to purchase. This seems to be worst in department store chains that utilize part-time help (who get shifted from department to department). I'm not saying you should avoid dealing with the big chainstore operations. Many times they will deliver the best price. What I do say is that you must be educated before you walk through the door so that you will know if the salesperson working with you has the necessary knowledge to properly handle the marketing of the product. If you know what you want before walking through the door, you will be able to spot and work with a knowledgeable salesperson. You will also be less likely to purchase a unit that has features you do not (and never will) need. This holds true when making any purchase, from clothing to automobiles. I call it the common-sense approach to buying.

The first step is to read this book and everything else you can on the subject. Know the difference between the two formats. Also make sure you know that these are the two emerging victors in a race that saw many of the competition fall by the wayside. Some of the fallen are still sitting on dealers' shelves and in their inventories. The prices may be attractive, but beware. There is no pre-recorded software available for these odd-format machines, and even blank tape for them is becoming hard to find. Don't get talked into a nonstandard "lead elephant" based on price. You will regret it.

I will assume that you will be purchasing either a Beta- or VHS-format machine. In both formats, there are machines with similar features and comparable price, so you should not make a hasty decision based on advertising alone. You should first ask yourself some simple questions and write down the answers. Let's run through a few of them.

1. Where will I do most of my recording? At home? Portable with a camera?

2. How much will I use the machine? Every day? A few times a week? Just when it's needed?

3. What will I be using it for? Recording a program while I am at work or traveling? Recording a program while I am watching another show? Making home movies? Business purposes?

4. Do I need such features as multiple-day programmability for delayed recording? High-speed scanning? Slow motion and/or stop action?

5. Referring back to Question 4, what material will I usually record? Nothing in particular? Sports events? Musical specials? Mainly for the playback of rented/purchased pre-recorded material?

6. Is this unit a necessity? A luxury? Am I buying it because I want and need it? Will it be a status symbol in my life? (Ego is a very important part of any major purchase. There is nothing wrong with achieving ego satisfaction through the possession of goods and other material items. This is not my feeling, rather that of many who are trained in the study of human character.)

7. Realistically, how much am I willing to spend for such a device, knowing that the initial purchase is just the beginning? (You will soon be purchasing blank tape or renting/purchasing pre-recorded material for viewing.) It is important to place a bottom-line dollar figure, taking variables such as the annual rate of inflation into account.

8. Which is most important in my purchase:
    A. Initial cost?
    B. Educated salesperson?
    C. After-purchase service?
    D. Compatibility with what most of my friends already have?

There are many other questions, but these eight give you a lot of insight into yourself and your particular needs. For example, suppose you are a TV sports addict. Many people are. You are buying a VCR because you don't want to miss a single moment of any game. Well, sports events are usually fairly long—several hours for some. Obviously you will want a unit capable of uninterrupted recording for that length of time. Well, with VHS-EP (SLP) you can go up to six hours, and Betamax III goes up to five hours. There are so many of those darn commercials. It would be nice if there were a way to bypass them when watching the tape. Ah yes—a machine with fast scanning. Both Betamax and VHS have this feature as an option. What about slow motion and stop action replay? Yep, both have that option available as well. Finally, what about remote control of every function in the machine (without any wires). You can get machines with that option as well. In fact, what you want is the very top of the line in either format. A VCR with all the bells and whistles—including multiday programmability to record any event you might otherwise miss.

On the other side of the coin is the person (or family) who goes away now and again and wants to record one or two shows for later viewing, or simply to record one program while watching another. For that person, all the bells and whistles are not a necessity. (Maybe saving a few bucks toward the kids education is.) The basic low-end machine with a mechanical tuner and 24-hour timer is more than sufficient. Saving the bucks is more important. They will go for the "advertised bargain." Somewhere in between, or maybe into one of these categories, each one of you contemplating the purchase of a video cassette recorder will fall.

There is another aspect of ownership to consider. A VCR is a very complex piece of electromechanical apparatus, possibly the most complex piece of machinery you will ever purchase. Service and routine maintenance must be done only by a trained and qualified technician. Who is going to repair your unit when it breaks down? You will want to consider the manufacturer's track record in relation to who performs their warranty service? How far away is the nearest service facility? How long does it usually take them to complete a repair? What kind of warranty comes with the machine? What does the warranty cover?

A person who services VCRs should have received training on the types of machines he services. Since many machines are very similar, it's not necessary that he attend every class and seminar given each time a new product is introduced unless that new product is a radical departure from what he is accustomed to. But, by the same token, just because a guy owns a radio and TV repair business does not make him qualified to service a VCR. Nor does a piece of paper on his wall that says he's authorized to service a given *"brand of product"* assure you that he is competent in videotape.

Some manufacturers require that every technician from a given service organization attend specialized training before certifying the service organization for VCRs. Others like Sears, Wards, RCA, J.C. Penney, and General Electric operate their own product support services. In some areas the service on these products may be performed by an outside contractor due to logistics, but in most of the densely populated areas you deal directly with the service organization operated by the manufacturer. This does not mean you can't find service elsewhere. It does mean you have a guaranteed avenue to obtain service if it becomes necessary. Finally, you will run into some off-brands and private-label merchandise that has no specific service structure. The best thing to do is ask the dealer about service. Some dealers do their own, some contract it out, and others send you to a manufacturer's authorized service facility (or one run directly by the manufacturer). Even this is no guarantee that the service you receive meets your particular standards, but it points you in the right direction.

If you find that you must deal with a service organization you know nothing about, you should at least ask if their technicians are trained on your particular brand of VCR. Look around the service center for documentation such as certificates from that manufacturer that state that the technicians are so trained. When I was in the retail service business I was very proud to display these certificates. Most technicians are. A technician must understand the intricacies of a system before he's given that certificate. (Over the years, I've attended many such classes and know this first hand.) To wrap it up, if a technician knows the basics of VHS and Betamax service, he can service just about any machine. If he specializes in one or two brands, he's likely to be very well versed on them.

While I can't give you any rule to follow, I will tell you that the majority of those servicing VCRs today are competent people. Like any other industry, this one has its share of quacks, but the very nature of the beast makes their careers short-lived. Also, the manufacturers have taken great pains to make sure their product receives good after-sale service if needed. After all, they hope you will come back to make another purchase from their line.

Purchasing a home video cassette recorder is not like going grocery shopping. It's more than just going over to the local electronics emporium and plunking down your hard-earned green stamps. It is knowing how to be a conscientious consumer—a person who takes the time to investigate all aspects of such a purchase including his or her own motivation. It is knowing how to shop wisely, but also knowing that the person who is making the sale is entitled to exist as well. That for him to stay around and succeed in business, he must be permitted to make a profit on that sale. (Take it from someone who knows.) The customer who comes into a store with the intent of haggling the proprietor into the ground is being "penny wise and pound foolish." I've watched the discount-buying craze destroy many product lines. I don't want to see this happen to the home video recorder. If it does, the person who will really suffer is the consumer.

Right now the standards of manufacture are very high in the home videotape field, almost as high (in many ways) as for the broadcaster's equipment. Yes, this does push the price up, but in the end it's product longevity that counts. If you spend a thousand dollars today on a high-quality piece of merchandise and it lasts you ten years with minimal after-sale service, that's a far better bargain that spending half that amount now, and twice as much keeping the product going during its normal lifespan. How long will a good quality home video recorder last? At this time there is no way to know. But remember what I said, at the beginning of this book, about my old Sony black-and-white open-reel recorder: That unit is still going strong and it was built in the 1960s. Yes, it's been well maintained over the years, but the service performed has never been more than routine maintenance. In my opinion, there is no way to put a price tag on quality.

Chapter 10

Getting the Most Out of Your VCR

Whether you've just purchased a home recorder, or have had one for many years, this chapter is important to read. It could save you a lot of time, money, and frustration. This is because a VCR—any VCR—performs only as well as the peripheral equipment it's connected to. This means that everything else in the TV signal chain must also be operating properly. If not, performance of the VCR will be degraded. Let's look into this step-by-step.

THE TV RECEIVER AND ANTENNA

How good is my TV set? Is it good enough for use with a VCR? The only rule of thumb here is that the television receiver should be in good repair, free of major defects, and able to accept a VCR. There are some TV sets, mainly pre-1975 tube-type that cannot handle a VCR without major modification. If you own such a set and are planning to purchase a VCR, it might be wise to first borrow a friend's VCR and try it. The most common problem encountered on such older sets is a "flag-waving" effect at the top of the screen (see Fig. 10-1). This is known as a *time-base error* between the VCR and the horizontal afc (automatic frequency control) circuit of the TV receiver. What's happening is that the horizontal afc of the TV set is slow in responding to these errors, so they appear on the screen as a flagging at the top of the screen. In some extreme cases, where an older set is in a state of disrepair, it might not be possible to achieve any horizontal lock-up at all.

What should you do if you find that your TV set falls into this category? You have several options depending on the vintage of the TV receiver and its manufacturer. Some sets can be modified to speed-up the action of the horizontal afc circuit, but this might be quite costly. (The price will depend

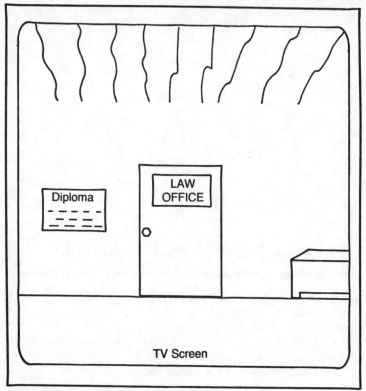

Fig. 10-1. If flag-waving effect appears at top of picture when VCR is connected to an older TV, the set may not be suitable for use with a VCR.

on the amount of parts, labor, and time involved.) You will have to place an exact dollar value on such a set, and then determine for yourself if modification is warranted. Perhaps the best avenue to follow is to purchase a new TV receiver of modern solidstate design to go along with your new VCR. If the TV receiver is approaching the ten-year mark, the chances are that it will need other repairs as well (especially if the picture tube is an original). In a case such as this, it's best to go with a new set, and keep the old one as a spare. Don't throw good money after bad, especially if the TV set has been trouble-prone since you bought it.

Even if you have an older TV set that does not exhibit a time-base error problem when operating from a VCR, there are a few other things you must be sure of. Can the TV set properly receive channels 3 and 4? Why is this important? Because most home VCRs use one of those two channels to feed the VCR output signal into the TV set. Usually the VCR manufacturer will recommend adjusting a slide switch on the VCR for whichever channel is unused in your area. This minimizes the chance of interference to the VCR playback. If you get your TV reception from a cable system, this may not hold much meaning, but if your reception is "off-air," from an antenna on the roof, then be sure you set the VCR output to that unused channel.

To determine if the tuner in your set is capable of receiving a channel not in use in your area means hooking up a borrowed VCR to the TV and playing back a pre-recorded tape. Make sure that both the VCR's output switch and the TV set are on the same channel. Put the VCR in "Play" and note the picture (if any) that appears on the screen. Next, take the fine-tuning control and run through its range both clockwise and counterclockwise. At some point you should get a very sharp picture, good sound, and good color. If not, double check to be sure that the VCR is outputting on the right channel and that the tape being used is good. If so, and you still can't fine tune a good picture and sound, then you can assume that there may be a problem with the set's tuner and it should be taken to a competent technician for repair or replacement. If the set is an "old clunker," it might be better to discard it in favor of a new one.

There are two other methods of checking a TV set's tuner for proper operation. Those of you who have *cable TV* probably have a local uhf station appearing on what would normally be an unused tuner position. For example, on the cable system that serves my area, channel 40 comes in on channel 3. If my set could not tune it in, then I would know I had a tuner problem. This test is obvious and requires no special knowledge. The simple rule here is: if it gets the channel on the cable system, it will do the same from a VCR.

The third way of finding out if the tuner in your set is operating properly is to have a TV serviceman come in and check it for you. He has instrumentation for this job, but it's the most expensive way to make the check.

It's not a bad idea to spend a few dollars *before* purchasing a VCR to have a professional check over your TV set. In fact, the older the set, the more important this becomes. There are some very subtle problems that develop with TV sets, problems that get worse gradually and are therefore not noticed. Possibly, after someone remarks at the poor quality of the picture, or he notices a difference between his set and that of someone else, the owner realizes his TV set has problems. A competent technician can spot such problems in an instant and effect repairs before a minor problem becomes an expensive major one.

Here's a short list of what should be checked.

■ Check picture tube with a good picture tube checker.

■ If it's a tube-type set, all tubes should be checked. Weak or defective ones should be replaced.

■ If it is a modular solidstate set, then module connections should be cleaned and all modules should be re-sealed to ensure proper contact with the motherboard. Any modules that exhibit obvious problems should be repaired or replaced.

■ Depending on the set, the following adjustments should be checked and re-adjusted as indicated:

—Vertical height
—Vertical linearity
—Color killer
—Horizontal width (linearity)
—Horizontal drive
—Static convergence (center resolution)
—Dynamic convergence (overall resolution)
—Overall screen purity
—Fine tuning, all channels.

■ If an outdoor antenna is used, it should be checked for corrosion on contacts, damage, and proper direction (toward station transmitting antenna). The downlead should also be checked for cracks, peeling insulation, and water contamination. If any such contamination is noted, replace the downlead, preferably with high-quality 75 ohm shielded coaxial cable. This may require the addition of matching transformers, but it's worth the expense.

If the antenna has withstood a half dozen severe winters, or an equal number of seasons exposed to salt air or bad smog, it's time to replace this antenna and downlead regardless of what a visual inspection might show. Trying to save a few dollars by salvaging an antenna that's a remnant of a bygone era is another way to be "penny wise and pound foolish." In an urban area, one need not spend a lot for a super-duper fringe antenna. The antenna used should be determined by what is considered optimum for your area. Consult a TV repairman on this one.

VCR MAINTENANCE

You've had your VCR for some time and suddenly the pictures you record are not as sharp as they used to be? Is your VCR broken? Not necessarily. Sharpness of a replayed picture is subjective, unless you possess the necessary equipment, test tapes, and knowledge to formulate a technical evaluation of a given machine. What you should do is compare the "original" with the "copy." That is, tune in a good-quality picture on the TV set. Then do the same on the VCR, and let the VCR record the program as you watch it. As soon as the show is over, rewind the tape and play it back, remembering that there is always some degradation between the air program and your recording. If the program playback looks as good as it has in the past, then forget it. There's nothing wrong. If the playback looks bad, then repeat the test again, this time using a new tape and with the recorder set to its fastest speed (Betamax II or VHS-SP). Again watch a show while recording it, and then immediately play back the tape. If the problem has gone away, chances are that the first tape was a fault. Tapes do have a definite lifespan and once past a hundred reuses, some tend to deteriorate very rapidly. If it turns out to be a bad tape, take that tape and throw it away. Bad tapes are notorious for causing head clogging problems.

Let's suppose the problem is quite severe. You have tried to clean up the picture using the VCR tracking control, but still the results, even at the highest record speed, are poor. At this point you need a competent VCR tech.

NEVER AND I REPEAT NEVER TAKE A VCR APART. The average consumer does not possess the necessary training to understand the electromechanical workings of these machines. Touch the wrong thing the wrong way and it could cost hundreds of dollars in unnecessary repairs. Don't think that reading this book or any other text will make you a competent VCR technician. That takes time and training.

What if it's only a dirty set of heads? Why can't I clean it myself with one of those cleaning tapes? Don't they work? Yes, cleaning tapes do work with certain provisos. While they can remove surface dirt, I've never seen one that will completely take care of a bad head clog. I consider cleaning tapes more a preventive maintenance tool than a repair item. When a bad head clog on the scanner does occur, it requires the use of certain chemicals to properly remove debris causing the problem. The most common is liquid Freon which is highly toxic and should never be kept at home. Some manufacturers recommend certain forms of alcohol to clean clogged heads, but not the drugstore variety, rather they are specialized cleaning solvents which are toxic. The application of these cleaners must be done properly. The head tips are very fragile and something as simple as rubbing a cloth across them in the wrong direction can damage them beyond repair.

The video heads and scanner assembly are not the only things that must be cleaned. When a clog has occurred, it usually indicates lack of preventive maintenance or a faulty tape (usually the latter). Tape going bad usually means oxide separation from the backing material after repeated use of the tape. Therefore, the oxide is not just deposited on the video heads, but on all surfaces of tape travel. This means that all surfaces that come in contact with the tape oxide must be thoroughly cleaned (as outlined in the manufacturer's service notes). On some machines this requires knowledge of how to fool a machine into threading itself without inserting a tape. It also may require a very expensive test tape for evaluation of the unit after cleaning. Again, this is where an experienced VCR technician is needed.

I am not against the use of cleaning tapes as long as you remember that they won't work miracles. Also, if you purchase one of these cleaning tapes, be sure it's nonabrasive.

What should I do if the repairman tells me it's going to be very expensive to repair my VCR? You really have only two options: First, if you trust the technician, then go ahead with the repair. Demand an itemized bill listing all parts replaced and the labor charged to replace *each* part. Also, be sure to have the old parts returned to you, unless the new parts are sold on an "exchange basis" at reduced cost to you. If this is the case, then it should be so noted on the bill.

Your other option is to "shop around" for another repairman, but remember that each technician who looks at your unit is entitled to charge you a minimum fee for such a service. Some states set this fee; other states leave it up to the marketplace to determine. Shopping around for service may eventually save you money but can be costly in the short term.

If tape doesn't last forever, what's the best tape to buy? Manufacturer's claims aside, there is no set rule. Rather than thinking in terms of brand name, it's more important to realize that tape has a finite lifespan. The more it is used, the sooner it must be replaced. Most of the name brand tapes are good. Some possibly better than others, but I will not single out a specific name and say that X is better than Y. I will say that you should stay with a brand of tape that's nationally known and in which you have confidence. Avoid private-label and bargain-priced tape cassettes. If a particular tape cassette starts to clog the heads on your machine regularly, toss it out—its life is over.

Do Beta-format cassettes last longer than VHS or vice versa? According to the material I have read, talks with tape manufacturers, and experienced dealers, the answer is "No." There is no set lifespan for any tape format. A bad machine can ruin a good tape even if the tape is new. Regular preventive maintenance is the best assurance of maximum tape and machine life.

How often should a VCR be checked on a preventive maintenance basis? This depends on the manufacturer of the machine. Each has his own ideas and if you follow the maintenance schedule outlined in most instruction books, you can't be too far wrong. My personal belief is that video heads should be cleaned regularly; after 50 hours of use, maximum. Most manufacturers believe in a far longer timespan between cleanings than that. I'm also a firm believer that rubberized parts such as drive belts and drive wheels should be changed on a regular basis, rather than waiting for the inevitable to happen while recording a show. There is a very simple way to know when belts need replacement. Take a new belt and the original from the machine, and drape both over a pencil or screwdriver. If the old belt seems obviously longer or larger, it's stretched and should be replaced with the new one. Also a visual inspection for wear, roughing, and cracks will tell if a belt or drive wheel is ready to give way. Remember that the parts are the cheapest part of the repair.

TAPE TECHNIQUES

My friend rented this tape and I tried to copy it but it won't record. What's wrong? Most of the pre-recorded tapes rented these days have a form of re-record protection built in. One system is called Copy Guard. There are others. The purpose of this is to prevent unauthorized duplication of pre-recorded tape, since most producers and distributors feel that doing so is a direct violation of copyright law.

There are certain "magic black boxes" on the market that are designed to beat the system, but they may be illegal in your state or locality. In essence, they are all variations on a broadcast theme called *time base correction*, but there is a difference. A broadcaster's Time Base Corrector (or TBC) costs more than $25,000 and is designed to optimize a playback signal for editing or air use. It's not a device designed so that someone can beat the copyright system. The home version of a TBC is far less elaborate and a lot less costly, but with apologies to those who derive a living from their manufacture, I can see no real advantage in them for the average home viewer. It's my opinion that the engineers designing home video recorders are well aware of the parameters necessary to reproduce a high-quality playback. They are not about to get themselves involved in helping the public become video pirates. My sincere belief is that with rental rates coming way down, the need to pirate a given film from a rental will soon be negated. If you **must** own a particular film, then go out and buy a copy. Don't get involved in pirating. Later on, I will get into the current legal aspects of home video in relation to this and other matters.

My machine gives a better picture at faster speed settings, but I can't record as much material. Is that the way it should work? Yes, the faster the tape forward speed, the better the recording and hence the recovered video on playback. It goes back to the old audio tape analogy where recording and playing back at 7½ ips gives better fidelity than 3¾ ips. (Refer to Chapter 4 for a more technical analysis.)

The special effects on my machine only work at one of its speeds. Is this right? It depends on the machine and I must advise you re-read your VCR instruction manual. In some VCRs, the special effects are designed to work best at a given speed, usually the longest play mode. Other units are designed so that the special effects (e.g., scan, stop action, and slow motion) work at all speeds. If you are in doubt, or if the instruction manual is not clear on this, I suggest you send a note to the manufacturer and ask him. Most manufacturers are glad to answer questions of this sort. If you include a self-addressed stamped envelope, chances are you will get an even quicker response.

What's the difference between a *home* VCR and an *industrial* VCR? While at some points in a given manufacturer's product lines, the two may overlap, the main differences are in the ruggedness of construction and the number of features found. Most home VCRs (excluding portables) are designed to be installed in one spot and left there. They will be used to record mainly from off-air signals and will not be subjected to very much user abuse. They are designed with the TV viewer in mind.

Industrial machines, on the other hand, are designed for non-air production and editing. Maybe a company wants to make a tape showing how their products are made, or an educational institution wants to distribute tapes of a training procedure. Since the recorders will be subjected to this specialized use, and will probably be moved about and used by various persons, the ruggedness of these machines is an important factor. Also, most industrial machines are single speed; the fastest speed for a given tape format. They may, or may not, include specialized editing circuitry, but most have provision for this on the rear panel. Most of these machines exhibit a few decibels (dB) better signal-to-noise ratio, but this would not even be noticed on an average consumer's TV set.

There is an exception to the above. In the latest portable ½-inch machines the main difference between the consumer and industrial versions is in signal-to-noise and the number of speeds. Most manufacturers seem to be moving toward the industrial design concept for the consumer portable thus realizing that in portable use, even a home recorder may find itself in strange environmental conditions.

Speaking of portable recorders, are they as good as regular home units? We are now into the "second generation" of home VCR and the answer to this one is yes. The original concept of a portable VCR seemed to be taking a regular home unit and splitting it into two parts. One case contained the tape transport, record/play electronics, and a battery. The other was a combined 24-hour tuner/timer unit, power supply, and battery charger. Although this basic concept still holds true, the weight has been significantly reduced. Today's portable VCR is less than half the weight of its predecessor. Advances in electromechanical design have permitted a reduction not only in weight, but also in size. At the same time, new head designs have improved overall picture quality.

The second-generation portable VCR can be purchased with multi-day, multi-event programmers to permit the same type of delayed recording as the single-unit machines. Most of the special-effect features are available as well. If your plans include making your own home movies on tape, then a portable system is the way to go.

VIDEO CAMERAS, VCR HEADS, AND VCR ACCESSORIES

What about cameras. What's the difference between the kind I use with my VCR and the ones the broadcasters use? There is a world of technical difference and a world of price difference. Most home cameras sell for under $1,000 and are designed to be used by people already familiar with home movie cameras. The idea is to make the transition from film to tape as painless and inexpensive as possible.

The average home camera has a single tri-electrode pickup tube designed to see the three basic colors used to make up a color picture in addition to the luminence (black-and-white) component of the picture. As stated earlier in this book, the tube is known as a vidicon. A number of variations on the basic pickup tube have been developed by various manufacturers. Hitachi's version is called a Saticon. There are others. The main variations on the basic theme are the overall sensitivity of the tube and its resolution. A simple rule of thumb—the more expensive the camera, the better its ability to see and reproduce a picture under adverse lighting conditions.

Most cameras used in commercial television, either in the studio or in the field, are *three-tube* design. That is they contain a separate tube for each of the three primary colors. One tube looks at red, another at green, and the third one at blue. So that each tube can get its share of the picture information, the lens does not focus directly onto the tubes; rather whatever image the lens gathers passes through a device called a *beam-splitter*. This will either be a "dichroic mirror system" or a "prism beam-splitter." Each tube feeds its own separate low-noise preamplifier and after processing, the three color signals are combined with the luminence (black-and-white) to form a composite color picture.

There are many reasons for the complexity. All have to do with overall sensitivity to light under varying conditions and broadcast standards as set forth by the FCC and other regulatory agencies worldwide. I guess the most important reason is that of overall resolution. Overall resolution is the quality of the picture perceived by the viewer on a subjective basis and the technological standard necessary to provide that picture. To that end, two cameras, one in the $1000 class and another costing many dollars more, are discussed in the next chapter.

You said that all home recorders use a rotary two-head system for recording and playing back video, but my friend just bought a unit that has four heads. Is that better than two? Almost every VHS manufacturer offers a four-head machine at the top of his product line. However, only two heads (working in pairs) are in use at a given time. At different record/playback speeds, different head gaps are required to obtain optimum performance—a question of overall reproduced resolution. To achieve this, two sets of heads are installed on the scanner (see Fig. 10-2). One set operates at the higher, or **SP**, speed, while the other pair are used in the slower long-play modes. The Japanese Victor Corporation pioneered this concept: in their single-scanner four-head design, 58.5 micron heads are used in the **SP** mode, while much smaller 19.5 micron heads are used in the extended play mode. Yes, this system does produce a better picture, but it also requires more electronics in the recorder with a corresponding higher price.

The four-head design has improved the picture quality of machines equipped with special effects such as slow motion, freeze frame (still), and scan. If you have a machine that offers these features at all operating

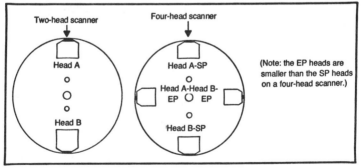

Fig. 10-2. Video scanners.

speeds, the chances are its a four-head unit. As of this writing, only VHS uses the four-head design, although rumors abound that Betamax format machines will soon use this concept as well.

Which is better, four head or two? That depends on your needs and how much you are willing to spend for a machine. If you are an average viewer who is interested in recording programs on a delayed basis for viewing at a later time, probably the extra bucks for four heads are not warranted. But, if you are a videophile or intend to use your machine for other purposes (e.g., evaluating sporting events or editing homemade videotaped movies) then the four-head design is definitely for you.

Won't the four-head machines be more troublesome than two-head machines? You should not expect any higher failure rate with a four-head system. I base this on my experience with commercial 1-inch open-reel production recorders which use six heads. While in television production work, changing these heads is a routine procedure, rarely must one be changed during normal operation of the equipment. Machines such as the ones I use at work are subjected to far more abuse than home VCRs receive, yet unexpected failures are virtually nonexistent. If you are planning the purchase of a deluxe four-head machine, I say go ahead, but remember the key to longevity is preventive maintenance on a regular basis.

What is a "Commercial Killer"? I can't tell you how many of my friends with VCRs ask me where to buy a magic box that will rid them of having to watch a recorded program riddled with commercial interruptions, nor can I tell you how much these magic boxes are despised by the folks on Madison Avenue.

A *commercial killer* is an electronic device that stops a VCR when a commercial break occurs and restarts it when the commercial break is over. It does this by sensing a time period in the video signal where *complete black* is present for a given time interval. This complete black is called *dc restoration*, and a commercial-killer unit simply samples this from the video output jack on the recorder. It also connects to the recorder's "Remote Pause" jack (see Fig. 10-3) and automatically stops the recorder for the first black period and restarts it in the second black period, after the commercial. These devices don't always work, as their users have found out. This is because it's conceivable for any of these units to be tricked by the TV station itself, especially if the station is running on "automation" where a computer is in control of the broadcast time schedule. It's possible for one of the two dc restoration periods to be bypassed and a recorder may then record the commercial and stop after it if the first period was deleted, or stop for the commercial break and then resume again at the start of the next commercial break if the second period was deleted by accident. Also, there are some forms of tape automation which "dip to black" between each advertisement. If that happens, the commercial killer will not know where it is and you only have a 50-50 chance of resuming the record mode at the start of the next program segment.

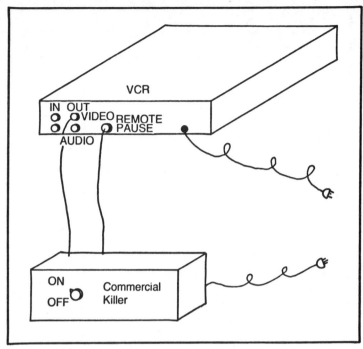

Fig. 10-3. Hooking up a "Commercial Killer" to a VCR. Remember, they don't always work!

More complex commercial-killer units are coming on the market that sample not only for "dip to black," but for other parameters as well. For those of you who may want one of these add-on toys, I suggest that you first check with someone else in your area who has one and find out how successfully the unit's operation is in regard to the broadcast standards in your area. By law the standards are nationwide, but from time to time men and machines will falter. I am not against such add-ons. They cannot harm a VCR in any way, but they are far from foolproof.

What's a "Video Enhancer" and what does it do? A *video enhancer* is another add-on black box which claims to clean up the picture when you are duplicating a tape from one machine to another (see Fig. 10-4). This process is called tape dubbing. Each time you re-record a tape from one machine to another, there will be a certain degradation of quality to both video and audio in the copy. (This copy is termed a *tape generation*.)

If you have two VCRs, here is an interesting experiment you can try. First record a program or other material at the machine's highest speed. Then hook the video output of that machine to the video input of the second machine. Do the same for audio into the respective jacks with the prescribed cables for each. (Never use audio cable for video recording—the losses are horrendous. Use video cables for video and audio cables for

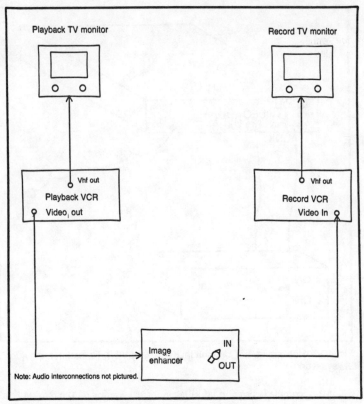

Note: Audio interconnections not pictured.

Fig. 10-4. Basic tape duplicator with "Image Enhancer."

audio, regardless of what anyone else might tell you.) Now play back the tape and record it on your second machine, at the highest speed in order to preserve as much resolution as possible. OK, now rewind and view that tape as compared to the original recording. Doesn't look as sharp as the first tape? You are viewing a second generation; the original being your *master* or first generation.

Now put the second-generation tape into the VCR and record a third generation. Look at it—by now the video degradation should be apparent. Carry it one or two steps further and it becomes quite annoying to watch the picture. Color becomes "blobbish" and outlines of figures become indistinct. Go a few more generations, and you probably will find an almost indistinguishable picture with poor audio to boot.

To help combat this problem, a myriad of devices, which all call themselves video enhancers, have been introduced in the consumer marketplace. I have tried a couple, and they seem to live up to their promises. The better units are those which feature *comb filters*, although even the simple *peakers* seem to help out in copying from machine to machine. Such

64

devices are used in the broadcast industry as well, but as you might expect these are far more complex than the consumer variety. In my business, video enhancement is a science unto itself. There are complex systems that can take an old movie, transfer it onto videotape, and the picture from the tape is better than the original film. A company called Image Transform, in Los Angeles, is one of those who pioneered image enhancement, and their work brought you the first high-resolution pictures from space.

On the consumer level, a video enhancer is only necessary if you have one of the following in mind: compiling an extensive tape library that's commercial-free, producing your own video films for wide distribution, or doing a lot of machine-to-machine editing. It's one of the few add-ons that I really like.

VIDEO EDITING

Speaking of editing, what is it anyhow? Editing is the process of assembling video and audio material in a progressive order. To understand the editing process, let's go back to my old friend, Super 8mm film. Suppose you have taken 60 minutes of film and it's sitting there on the shelf. In that 60 minutes is actually a five-year span of your child's birthday parties. One day you decide you want it all on one reel so that you don't have to change film every five minutes. You know how frustrating that is. So you go out and buy a film-splicing machine, some splicing tabs (or glue for you purists), and some empty reels. Then you run the first film, and when it ends you splice the start of the second film to it. Then the third, fourth, and so on. Soon you find that you are only half finished, but there is no more room on the take-up reel. Ah! Most Super 8mm projectors only handle a 400-foot film reel that runs 30 minutes. So you rewind the film you have already spliced together onto the 400-foot reel and begin the process of splicing the remainder of your 60 minutes onto the second reel. Lucky you purchased more than one.

A few hours later, you are ready to sit back, relax, and enjoy the fruits of your labor. You thread the projector and settle back into your easy chair. You are satisfied. What you do not realize is that you have performed a very basic form of something known as *assembly editing*. You have *assembled* your film into the order in which you want to view it.

It's now a few months later. You've seen the film several times and noted your mistakes. If only there were some way to cut them out, the film would look so much better. So, you get ambitious again, trundle down to the local camera store, and purchase a super 8mm editing machine. This is nothing more than a small rear-projection viewer and a couple of hand winders. You set up shop on the kitchen table and slowly look at the film on the editing machine. When you reach the point where "Junior" kicked you in the shin, causing you to photograph the ceiling for the next 15 seconds, you simply remove and discard that section of the film. After a while, you note that there is a lot of film going into the scrap heap. "Could I be that unprofessional?" you think to yourself. Fear not. In the educational movie business, it's common to shoot "20 to 1." That is; for every minute of finished product on the screen, 19 minutes of film have been discarded. By the time you're finished, you have all the film you want to keep on a single reel. You view it, and again are very pleased. Things seem to move at a better pace and it just looks better. What you have done is *edited for continuity*.

There is one more step. Suddenly you realize it would be nice to have titles and maybe some other background material in your film. You go out and get some film, and tripod, and a titling set; photograph titles and get that reel developed. You also have moved twice in that five years, so it would be nice to include some background footage of each home. Out comes the editor, splicing machine, and splicing tape. Then a thought occurs to you, "It just fits on one reel now, but if I add the new material it won't. The answer is to sit down and watch the movie again, making notes of what comes after what. List things scene by scene." Hmmm—do I really need four minutes of Judy opening up her presents? Maybe I can shorten that to two or three minutes. Yes—just after she hugs grandma—that's a good point to cut—"Congratulations, you have discovered another coveted prize of the broadcaster and film maker—the *outcue point*.

So you start in on the tedious process of cutting "this" out to make way for "that." A title goes at the very front of the film, followed by a ten-second glance at the first house, then a title saying "Judy-Age 1," then the party. True, you had to delete some party film to put in these other shots, but they really add to the overall story. So you insert new material in place of the original film. Congratulations again, you've now discovered the third form of editing; that of *insert editing*.

The same basic procedures hold true for video editing on tape, with one important difference. *Videotape editing is totally electronic. You must never even attempt to splice videotape. Splices can cause severe head damage to the VCR's scanner assembly.*

If it's electronic, what equipment do I need to do basic video editing? Actually, all you need are two VCRs that are designed to be used in an editing configuration, some cables for interconnecting the two, and a pair of TV sets to be used as monitors, and a plentiful supply of tape that's been pre-recorded to "black." Later in this text, I will explain exactly how videotape editing is performed and will discuss some additional equipment that will make your editing job a lot easier.

AUDIO AND MICROPHONES

What about enhancing audio quality? Are there gadgets for that job too? Yes, there are. In fact the two most common are units that you may already have as part of your stereo system. They are a multichannel audio mixer (some preamplifiers have this feature) and a graphic equalizer. Used properly, they can help "sweeten" audio quality in editing. A third device called a noise reduction system such as Dolby or DBX is also nice, but with the limited audio passband of most home VCRs it's not essential. Even in commercial television production, these systems are only now starting to make any real entry—based on the projected future of "Stereo Home TV" and digital audio on TV.

Speaking of stereo TV sound, why aren't home video recorders equipped for stereo sound? Until stereo telecasting becomes a reality on national and international levels; home machines will probably retain the monaural record mode of operation. It is possible to place two separate discrete audio tracks on ½-inch tape along with video and control information. Many industrial recorders now on the market feature such features as options or as part of the initial design. The SLO-320 VCR described earlier. (Sony Betamax I format) is such a machine. But, if you are waiting for stereo to become a part of the home VCR before purchasing a machine, you may be in for a very long wait.

My buddy has a camera and a portable VCR, but the audio quality is poor. Why is that? Probably because he has a good recorder and camera that are both working properly, but he is relying on the camera's built-in microphone for all his sound. Such microphones are a compromise at best. They are subject to picking up extraneous noise generated by the camera operator himself. In some situations they may be adequate, but there is no substitute for using the proper microphone for the shooting session. A glance at most home recorders and home cameras shows that both usually have a jack marked "Mic In." If you are recording any distance from the subject, it pays to use an external microphone. There are two basic types I recommend for the home movie/tape maker. The handheld *dynamic* microphone is best for use when your subject knows the proper way to use a microphone. Taping someone inexperienced in speaking before a microphone and camera, there is no substitute for the "tie-clip" *electret condenser* microphone. They are tiny, inconspicuous, and work amazingly well. I do not recommend their use on young children, since they get fidgety, and a microphone becomes an interesting toy. Here, a third type of microphone known as a *unidirectional* or *shotgun* microphone mounted on a stand and aimed toward the subject is most appropriate.

What about cost? In my case, I have spent several hundred dollars on a collection of different microphones for use in various environments. I use Shure SM-61s as hand microphones and Sony ECM-50Ps for tie-clip service. I rent unidirectional shotgun microphones as the need arises—usually something like a Senheisser 805 or 815. You need not go to the extremes I have in microphone selection. I had to do it because it's a part of my livelihood. Otherwise, I probably would have visited my local Radio Shack store and purchased some of their microphones. One thing I have found about Radio Shack/Tandy Corporation is that their products perform to specification. No, they are not as rugged as equipment meant for broadcast field production or news gathering, but the audio quality from their microphones is as good as they claim. In fact, I use a number of their under-$25 microphones for radio news recording both in my studio at home and at a facility in Hollywood. The main thing to remember in purchasing an external microphone to use with your recorder, from Radio Shack or anywhere else, is that the microphone output impedance must match the microphone input impedance of your particular recorder (usually on the order of 200 to 600 ohms, although some recorders do require use of higher impedance microphones). Before purchasing a microphone, it's best to consult your VCR owner's manual to find out if your machine will accept an external microphone, and, if so, what the impedance is. One other note: all home recorders are *unbalanced input/output* on all audio lines. Be sure that the microphone you get is wired for use in this configuration.

Excessively long microphone lines can cause noticable hum in the recorded audio. If you have to go less than 20 feet, you will probably have no problem. Get much past 30 feet, and the hum problem may get noticable. There are at least two ways around this. The simplest method is the use of a narrow beamwidth directional shotgun microphone, although unwanted reverberation could be a problem. The other method is to buy or rent a microphone designed for use with low-impedance balanced line and use a matching transformer directly at the microphone input to the VCR. With balanced audio line, you can go a hundred feet or more without noticable hum pickup, and the matching transformer will convert the balanced line to unbalanced at the proper impedance for your VCR's input. Finally, if you have to go several hundred feet between camera and subject, then you can rent wireless microphones which will allow you total freedom from wired operation. A wireless microphone is exactly what the name implies: It's a microphone that's attached to a miniature FM transmitter. (Some are even built right into the microphone case.) The other end of the circuit is a receiver that plugs directly into the VCR's microphone jack. Both are battery-powered and run for hours. Some manufacturers who market this type of microphone are Swintek, Cetec-Vega, Edcore, and H.M.E. Again, these are meant for use in commercial broadcasting and production, but knowing about them can be helpful to the neophyte home video movie maker.

If you wonder why I am dwelling so heavily on the audio aspect of home video recording, it's because I strongly believe that the best pictures are worthless if the audio quality that goes with them sounds like it was recorded with a tin can and a piece of string as a microphone. I'll close by simply saying, "A picture that's worth a thousand words can easily be spoiled by rotten audio." Keep audio as important an objective as video in any recording you make. It's worth that extra effort.

FILM TRANSFER TO VIDEOTAPE

What about my collection of home movies, can I put them on tape? Yes, and it's as easy these days as a visit to your local camera store. A few years ago, the idea of film-to-tape for home video was pioneered by the Fotomat chain. Their success soon convinced other companies to offer similar services. For a set fee, they will transfer your 8mm or Super 8mm film onto either Betamax II or VHS-SP. You can do it yourself if you own a camera, but the quality of having it done by a professional organization is worth what amounts to a rather minimal expense. Here are the reasons.

First, the technical end of it: Most 8mm and Super 8mm film is projected at 18 frames per second. Super 8 sound is projected at 24 frames per second, as is 16mm sound film. What this means is that every second you view a Super 8 movie, you are seeing 18 or 24 different pictures per second. But in television you are viewing 30 frames per second, or thirty different pictures. If you simply aimed your camera at a screen and started both your projector and VCR, yes, you would get a picture, but the difference between the number of film frames and TV frames would appear as a bar rolling up the picture. The term for this is *scan rate difference*. To compensate for this, special equipment is used to project film onto TV. The device is called a *film chain*, and is nothing more than a projector with a special shutter that makes the camera (TV) think it's seeing 30 pictures each second. Admittedly this is an oversimplification of a rather complex process that involves electronic compensation and enhancement techniques, but it is the rudimentary basis of what is called *Tele-Cine*. Anytime you see a movie on television, it's either being projected on-air or from videotape using a Tele-Cine film chain.

One suggestion: Before taking your film to a transfer service, it might be wise to at least do a basic assembly edit on it. This will make it a lot easier if you intend to re-edit as previously outlined. It also insures that the film will be transferred to tape in the sequence you desire, even if you plan to do nothing more than view it. And one final note. When you view your tape, you may notice that parts of either side of the picture are missing (same for the top and bottom). That's because the *aspect ratio* of film is different from that of television. Keep this in mind when shooting film you intend to transfer to video, and try to keep all important action in the center of the screen. Rather than bore you with numbers, the best suggestion is to have one of your films transferred to tape and then view it alongside the film itself. The aspect ratio differential will become quite obvious. This is a case where experience will be the best teacher.

What about putting my color slides onto videotape. Can this also be done? No problem. In fact, it's far less of a problem than with motion picture film. Since we are dealing with a static field rather than a series of frames in motion, transferring slides to videotape is a far less complex process. In fact, I have done this many times using relatively inexpensive TV cameras and a slide projector. There are two ways you can do this yourself. One way is safe, but will give varying results, while the other yields excellent results but should only be attempted by someone who understands how to modify a slide projector so as to vary the intensity of the projection lamp.

The safe method (see Fig. 10-5) is to set up your slide projector and TV camera side by side, focused on a small neutral-gray screen a few feet away (just far enough to give a sharply-focused picture about 10 to 12 inches wide). Do not use a regular projection screen for this. Both glass beaded and lenticular screens are far too reflective. You want a dull, nonreflective surface such as gray construction paper. Now, turn on your projector, focus

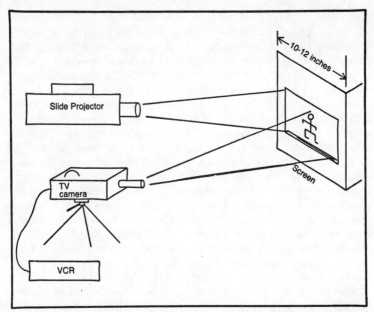

Fig. 10-5. The "safe" way to put color slides on tape is to set up your camera and projector side by side, and tape image from a small neutral-gray screen.

for a sharp outline of the film gate, and zoom in on that lit spot until it completely fills your viewfinder. Next, perform the *white balance* adjustments (including proper filter selection) as outlined in the owner's guidebook for your particular camera.

Once this is done, project your first slide and focus as sharply as possible. Now, go to full zoom on your camera lens and focus it as sharply as possible. Once completed, zoom back until the image just fills the viewfinder, top and bottom. If necessary adjust the tilt on your tripod to center the image vertically, and adjust the pan for horizontal centering. If you have a choice of automatic versus manual iris control on your camera, use the latter. Adjust the lens f-stop to a point where the brightest "whites" just start to get shiny. (This is the white-clipping point.) When you are satisfied with your adjustments, record a few slides at about five-second intervals and then replay the tape on your TV set or monitor. Adjust the color and tint controls to get as close an image to the original as you can. If you are nowhere near the original at any point in the tint range, then white-balance the camera again. Also try taking out the filter. There is no set rule as different cameras perform this task in different ways. Experiment until you find a happy medium. Remember, that you probably won't be able to exactly duplicate things as they look on the screen, but come as close as you can. When you are satisfied, choose a time length that each slide should run. (Some projectors can change slides automatically at preset intervals. Now is the time to make use of this feature.) Take a stop watch (or any watch with a sweep second hand), start your VCR, go to Slide 1, wait the alotted time, then to Slide 2, and so on. You can also liven things up a bit by using the camera's Zoom to look at specific areas of interest, but always remember to zoom back out slowly before going to the next slide.

The other method is to literally construct your own "Tele-Cine slide chain" by placing a Variac transformer in series with only the projection lamp, exchanging the lens on the projector for one with a long focal length and reversing the slides in their carrier so that right is left and vice versa. Also, you need a camera that has a detachable lens and manual electronic gain (or a fast-response automatic-gain circuit). Here's how this one works.

Mount the camera on a tripod with the lens removed. Dim the lights in the room enough so that you can still see what you are doing, and make sure that the camera is not pointed directly at any room lamp. (The reason I do not recommend this method to anyone other than an experienced videographer is that one wrong move can cause severe damage to the camera pickup tune. I am detailing the procedure but take no liability, if you damage your equipment trying to duplicate what is described herein. In other words, you are on your own!) If camera gain is manual, start just above minimum.

Connect the TV camera to the VCR and the VCR to a TV set or (preferably) a color monitor. Adjust the Variac transformer to minimum, turn on the projector and slowly advance the Variac until some illumination is noted on the screen of your monitor. Now, focus the projector onto the target of the vidicon tube until you can see the outline of the film gate quite sharply. Advance the setting of the Variac transformer a bit if necessary. Once the outline is sharp, do a quick white-balance on the camera. Project a color slide and view it on the monitor. Re-focus until the picture is sharp, clear, and crisp. Remember, you are projecting an image that's only about ¾-inch wide and focus will be critical. Still with a dim image, adjust camera centering and distance from camera to projector until you achieve the aspect ratio you desire. Now, slowly, ever so slowly, advance the Variac transformer and camera's gain until the picture on your monitor looks good. Again, whites should barely be shiny on cameras that have no automatic gain or those with automatic gain turned off. Keep the light source only as bright as necessary, because excess illumination can destroy a pickup tube faster than you can read this sentence. From this point on, never expose the camera to the projector light without a slide in the machine, but do project a slide with a lot of white and white-balance the system again. Again it will be "cut and try." Once you are satisfied, record your slides as outlined earlier.

When finished, immediately turn power off to the camera and reinstall the lens and lens cap—protect your camera. In either method, both narration and music can be recorded simultaneously or, if your VCR has an "audio dub" feature, they can be added at a later time. The latter method will give better resolution, but the creative ability to use the camera's zoom feature is lost. Keep this in mind when making transfers.

While it is possible to get good results using either method just described, it's also important to remember that you will be spending a lot of time on such a project. These days, it is far simpler to take your slides (or movie film) to a transfer service and have them do it for you. The cost is only a few dollars more than that of a blank tape cassette, and they guarantee the results.

TAPING FROM CABLE-TV AND INTERFERENCE

My VCR has an all-station tuner, and I can get the movie station on the cable but the picture is all scrambled. Is there something wrong with my VCR? No, your VCR is working properly. Most cable companies purposely scramble both the picture and sound on what they call their *premium channels* for the sole purpose of preventing people not paying for the service from receiving it. Yes, your VCR and possibly your TV set may tune these cable *mid-band* and *super-band* channels, but neither has the required decoder built in. Laws in many states preclude private ownership of such devices although "pirate decoders" are known to exist.

To complicate matters further, many cable TV companies will flatly refuse to connect the output of their *premium channels* to a videocassette recorder in your home, because doing so violates copyright law. In fact, a ruling by a three-judge Federal Appellate court in California (October, 1981) states that it is illegal to record off-air any copyright material, including regular TV. It's expected that copyright laws will eventually be amended to permit the latter, and with the number of VCRs already in the consumer marketplace, it's doubtful that anyone will ever be cited for recording his favorite TV program, but the recording of subscription pay TV is another story. In this, you are totally on your own. More on this later.

I have a "CBer" nearby and he interferes with my VCR and TV. What can I do about it? First, don't bother writing the FCC. They can do nothing to help you other than send you a form letter and a booklet on how to minimize your problem. The fact is that it's your problem and not that of the transmitting station. If you are on cable TV, ask your cable company for assistance. They have equipment and personnel trained in handling these problems. If you use your own antenna, have a competent TV repairman install an R.L. Drake "High-Pass" filter on your set. There are two models: the TV-300HP for use on flat 300 ohm twinlead, and the TV-75-HP for use on 75 ohm coaxial cable. Proper installation means that the filter must be properly grounded to the VCR or TV set. Many of today's TV sets are of transformerless design, and improper installation can lead to a serious shock hazard. AGAIN: *Refer such installation to a qualified service technician.* Experimentation by "yours truly" using many filter designs has shown that those made by R.L. Drake are the most effective, but not foolproof. If the problem still exists, check your antenna and downlead for corrosion, decay, and damage. If the antenna is defective or the downlead contaminated or aged, replace both using a high-quality antenna and shielded coaxial downlead.

If all else fails, then it's time to approach the source of the interference problem. Never, approach such a person with a militant attitude on your part. That's the quick way to get nowhere. Remember a CBer or ham radio operator has the same right to operate his station as you have to enjoy your TV and VCR. Even more, in that he or she has had to obtain a license from the FCC to operate a station. In the case of a radio amateur, he or she has had to pass a test in order to obtain that license and must abide by Commission rules and regulations. The FCC has set specific emission standards for both CB and amateur equipment. If the equipment meets those specifications, then he is in the right.

The best way to handle such a problem is on a friendly basis. Approach the person you suspect is the problem-causing source and request his or her help. A simple, "Hi there, I'm your neighbor and I think we have a problem." will do a lot more for your cause than saying "you'd better get that (expletive deleted) thing off my TV, or else." I have to ask—"or else what?" Sure you can hire an attorney and seek injunctions, but what will that get for you other than legal fees paid out and an enemy for life? The key to solving any interference problem is mutual cooperation on a friendly, neighborly basis. A few years ago, I served on the Television Interference Committee of a large Los Angeles amateur radio club. Many times we handled complaints referred to us from the local FCC office. Most of the time it was a matter of simply getting the two parties talking with one another and then rendering whatever technical aid we could, short of making repairs to a defective TV set or stereo system. Most of the time we were also successful in either minimizing or totally eliminating the interference problem. The key to our success was simply instilling confidence in all parties that a solution could be found, and then taking a professional approach in that direction. I won't tell you that this approach is foolproof, but it's no secret that you catch more flies with honey than with salt.

There are cases where no solution is possible. I've seen this a number of times. For years, radio amateurs through their national organization (the American Radio Relay League) and various CB radio groups have been trying to get Congress to pass legislation which would permit the FCC to set minimum rfi/tvi susceptibility standards for all home-entertainment electronics. Each year, the powerful Electronics Industry Association lobby does all it can to stand in the way of such legislation. As this book is being written, another bill (Senate Bill number 929, sponsored by U.S. Senator Barry M. Goldwater) is in hearing, and again the EIA is adamantly opposed to it, stating that the manufacturers are in the process of solving the problem. This is the very same "story" that the EIA has been using for years, and I have yet to see any major improvement in relation to radio frequency interference susceptibility of consumer electronics.

Nor are CB and amateur radio stations the only sources of these problems any longer. The newest culprit is the personal computer, and their numbers are growing by leaps and bounds. I've known of cases where such devices have made televiewing impossible for those in a household as well as for the neighbors. Unless stringent standards for rfi susceptibility are enacted, or the manufacturers get with it as they have promised, things will get a lot worse before they get better. Rfi is a major problem of the 1980s, and I know of no sure, fast cure.

What's next on the video horizon? Here are my personal predictions for the future of home video. Video recorders, which are still a luxury for many, will become almost a necessity to daily life. They will become a teaching tool, a learning aid, and a part of our day-to-day life. The same will hold true for the personal computer until it gets to a point where life without one may become impossible. They and other video-related products will become a part of our lifestyle, and we will wonder a few years from now how we were able to exist before their arrival on the scene.

Cable TV will make immense inroads as well. It will not replace commercial TV but rather it will augment it and give the televiewer a wider choice of entertainment. By the same token, I predict that Federal legislation will be enacted to protect cable telecasters and other "over-the-air premium programmers" from theft of service, but that the question of legality of home videorecording of such premium programming will be in the courts for years to come. It may never be settled. Eventually, the premium telecasters will realize what I have believed all along. The answer to this will come from the public and not the courts.

Direct-to-home satellite telecasting will also make inroads. *DBS*, as it's called in the trade, will allow a small rooftop antenna and special converter to pick up programming from special high-power satellites in geosynchronous orbit. This too will be of the "premium subscription" variety, but I am willing to bet that those planning such a service will learn from past mistakes and develop some form of "piracy proof" encoding/decoding system, something complex enough to make it uneconomical for the consumer to own or build. Also, with the advent of DBS, I see an end to the early 1980s craze toward giant home earth stations to receive the current generation of communication satellites. Let's face it, a 10- to 12-meter dish antenna in the backyard is not especially beautiful to look at. A 2- or 3-foot dish on the roof is far less an eyesore and far less expensive.

Finally, I see a great future for two-way interactive cable systems. There may come a day when we will be able to do all of our household shopping chores, banking, and even vote from our living room chair. The future of home video is limitless.

Chapter 11

The Video Camera

I said that we would delve deeply into the video camera in this chapter. In doing so, I am making two assumptions: first, that you will be interested in learning about this fantastic device, and secondly that someday you may own one. Let's start by establishing a standard for comparison. For this purpose, I have chosen a camera well known in the broadcast industry, the Ikegami HL-79D. Many video journalists will tell you it's the "Rolls Royce" of electronic news gathering and electronic field production. I've used Ikegami cameras for years, and I have to agree. The Ikegami is not the only camera in this class, but for some reason it and the RCA TK-76 have become broadcast industry standards. While you may say that comparing the average $1000 (or less) home camera to a unit like this is like trying to compare apples with oranges. I have to state that from my point of view the closer a home camera can get to a professional camera in specifications, the better the home camera will perform. In case you are wondering how much a camera such as the HL-79 costs, a close buddy just sold his, used, for $38,000. I shudder to think of telling you the retail list price. You have probably seen cameras like these covering news and sporting events. For example, they were used by ABC in its coverage of the 1980 Winter Olympics from Lake Placid. Need I say more? Figure 11-1 is a listing of the camera's technical specifications that we will use for comparison. First though, some explanation of terminology is in order.

PROFESSIONAL STANDARDS

Lets go through Fig. 11-1 step by step. First I will deal with the "rating" column. I'll start with Section 1. Input Signal by saying that Item Number 1 (External Sync Signal) is of no interest to the average videophile, but it is

Rating

1. **Input Signal**
   1) External Sync Signal: VBS/BBS 1V(p-p) positive, 75 ohms
   2) Return Signal: VBS/VS 1V(p-p) positive, 75 ohms
   3) Program Audio: −60 dBm, high impedance
   4) Tally: 24V dc or contact closure
   5) Remote Control:

2. **Output Signal**
   1) Composite Signal: VB/VBS 0.7V/1V(p-p) positive, 75 ohms, 2 outputs
   2) Monitor Output: VBS 1V(p-p) positive, 75 ohms, 1 output
   3) R-G-B Video: R,G,B 0.7V(p-p) positive, 75 ohms
   4) Program Audio: −20 dBm, 600 ohms
   5) VTR Control: Start/Stop
   6) Intercom: 2-wire system

3. **Pick up Tube:** 2/3 inch Low Capacity Diode gun Plumbicons®
   2/3 inch Diode gun Plumbicons® XQ-2427
   2/3 inch Broadcast Quality Plumbicons® XQ-1427
   2/3 inch Broadcast Quality Saticons® H-8398

4. **Viewfinder:** 1.5-inch high resolution Viewfinder

5. **Filter**

ND	0	0.6	CAP
COLOR	3000°K	4200°K	5600°K

6. **Power:** Ac 100V/115V, 50/60 Hz
   Dc 12V, 2.5A approx.

7. **Ambient Temperature:** Camera Head −20°C ~ +50°C (−4°F ~ 122°F)

8. **Lens Mount:** Bayonet mount

9. **Weight:** 6.7 Kgs (including 1.5 inch viewfinder)

Fig. 11-1. Technical specifications of HL-79D, professional video camera (courtesy Ikegami Electronics, Inc.).

extremely important to the broadcaster. This is because the videophile will rarely use more than one camera at a time, while in broadcasting multiple-camera installations are commonplace. To make sure that the picture won't roll or tear when a director switches from one camera to another, all cameras in a given studio installation are fed from a single source of synchronizing, or *sync signal*—this is usually known as a "house sync generator". In fact, in broadcasting and production, all equipment that

Performance

1. **Frequency Response:** (100kHz reference)	1) 50Hz ~ 5.5MHz: +1 dB/−2 dB
	2) Less than 50Hz, and more than 5.5 MHz; falling down
2. **Resolution:**	(with RETMA standard resolution pattern at 2,000 lux)
1) Center	More than 600 TV lines
2) Corners	More than 500 TV lines
3 **Geometric Distortion:** Overall picture area	Less than 1.5%
4. **Registration:** *	
1) Zone No. 1	Less than 0.1% of the picture height (within circle having a diameter equal to 80% or picture height)
2) Zone No. 2	Less than 0.2% of the picture height (within circle having a diameter equal to picture width)
3) Zone No. 3	Less than 0.5% of the picture height (outside of Zone No. 2)
5. **Sensitivity:**	Standard sensitivity: 2,000 lux, F5
	Mini. illumination: 20 lux. F1.4 (+18 dB Video Gain up)
6. **Signal-to-noise ratio:**	57 dB (using Low Capacity Diode gun Plumbicons®, with Gamma, Detail correction all off, Band Width: 4.5 MHz)
7. **Stability:**	
1) Supply Voltage Fluctuation ac:	±5% of rated voltage
dc:	11 V ~ 16V
2) Input Signal Fluctuation	Stable operation in the range of −6 dB ~ +3 dB.
3) Ambient Temperature:	When the ambient temperature varies ±10°C (±18°F) of the setup temperature in the range of 0°C ~ 40°C (32°F ~ 104°F), specifications are satisfied without readjustment.

* Plumbicon® Registered Trade Mark of N.V. PHILIPS
* Saticon®   Registered Trade Mark of HITACHI LTD.

All specifications are subject to change without prior notice.

requires sync is tied to this common source, as standard practice.

Most home cameras have no provision for insertion of external sync. The camera internally generates its own. All professional cameras have this feature as well, but it's important that they can also be synchronized to a common source. For this reason, the broadcaster's camera has the added provision that enables the internal sync signal to be shut off, allowing the camera to operate from external sync control.

Item Number 2 is called *return signal*. It's also called *return video* and is nothing more than externally switching the viewfinder on the camera to another video source. As a comparison, some home cameras feature electronic viewfinders that give the owner the ability to view his recording on the spot simply by rewinding the tape and playing it back. The viewfinder on the camera also can be used as a mini-monitor for the VCR.

Item Number 3 is *program audio*. An earphone can be plugged into most home cameras to monitor record and playback audio. In the broadcast studio, audio is picked up by a number of microphones separate from the camera and processed as a separate entity. Even in the field, where reporters for news shows are using a handheld microphone the audio is kept separate. This is called program audio and in many cases it's helpful for the cameraman to hear it. This is especially true in one-man electronics newsgathering situations where one person is the camera operator, audio engineer, and recorder operator. This situation is not much different from you recording in your home, and for this reason video cameras are equipped to monitor program audio.

The term *tally*, Item Number 4, refers simply to a lamp on the camera that tells the cameraman and those being photographed that the camera is in operation. Some home cameras also have this feature. Item Number 5, *remote control* can have many meanings, but here it means that the camera's many controls can be operated externally by a device known as a *camera control unit*, or CCU. The closest that the typical videophile will ever come to a CCU is, perhaps, an external power supply for his camera, but in the studio, external control by a trained video operator is routine. Remote control at a CCU allows cameras to be matched for chroma level, hue (tint), and other factors. It also lets someone other than the cameraman worry about the correct lens setting to get the proper amount of light into the camera. All the cameraman has to do is set up the shot the way the Director calls and keep the camera in focus. Sounds easy? No way—not with the pressure of today's TV production business. A good cameraman is one who can literally be doing five things at once. Not just the foregoing two, but he must also be able to save a shot should talent miss a mark, or some other unexpected happening take place. That's why good cameramen are well paid. By the way, the procedure of walking the cameras and talent through a show, scene by scene, a logging each position is called *blocking*. It's an integral part of the rehearsal process for any show and is practiced until everyone knows where to be at any given moment. It's a key to successful production and something that you can do if you should undertake a project to put a mini-movie on tape, yourself.

Section 2 (in Fig. 11-1) is a listing of output signals available from the camera. Item Number 1 *composite signal* (sometimes called composite video) is a standard for home and professional cameras alike. Unless your camera is an oddball, the output will be that rating shown in Fig. 11-3, or within a few percent of it. If your camera uses an electronic viewfinder, then it too requires a composite video signal from the camera to operate. Even if the viewfinder is built into the camera, it's actually a separate device, and this explains Item Number 2. As to Item Number 3 *R-G-B video*, these are outputs from each separate tube in a three-tube camera. R is red, G is green, and B is blue. This is used in the camera registration process, and is something that you need not be concerned with unless yours is a three-tube camera. Most home cameras use a single *tri-electrode* pickup tube, but three-tube cameras require special alignment to make the tubes work in unison to deliver a proper black-and-white (monochrome) picture. This can be done at the camera itself or remotely using a CCU.

Something you won't find on even the best home camera is Item Number 6, an *intercom*. That's used only in broadcasting production work and allows the director to talk with the cameraman. Item Number 5 you will find on even the most inexpensive home camera—a remote *Start-Stop switch* for your VCR.

Section 3 (Fig. 11-1) is really something we can use for camera comparison. When you purchase a home video camera, you have no say over what type of pickup tube it will come with. In fact, until you get into cameras costing $6000 to $10,000 you have little to say. This does not hold true above that point. As you can see, the HL-79D, and most other broadcast-quality cameras, offer pickup tube options. The basic tubes are *broadcast quality* Saticons™ (Hitachi, Ltd.), while higher-resolution tubes are available at higher cost. It's interesting to note that Saticon tubes are very popular in single-tube home video cameras. Without getting overly technical, the Saticon tube is Hitachi's modification to the basic vidicon discussed earlier in this book. Other manufacturers have specialized trade names for their tubes as well.

There is another important difference between the pickup tube in a single-tube camera and one in a three-tube configuration. (Refer to Fig. 11-2 and Fig. 11-3). Figure 11-2 shows a basic single-tube camera configuration. Regardless of the tube designation, here's what happens: All light passes through the lens and is focused on the target of the pickup tube. As described earlier, the tube looks for differences in light intensity and varies its output level at any given instant accordingly. In addition it looks for chrominence (color information) and converts that into an electronic signal as well. The tube actually sees only three prime colors—red, green, and blue. There is an output terminal for each color signal which is then amplified and further

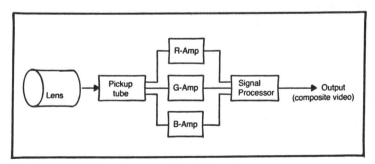

Fig. 11-2. Single-tube color camera (basic).

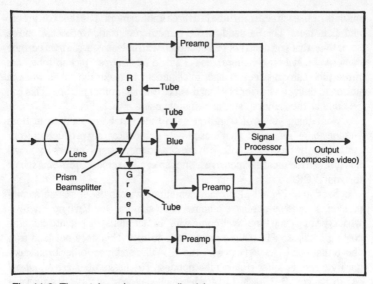

Fig. 11-3. Three-tube color camera (basic).

processed into a composite video/chroma signal. That's your basic home camera.

Figure 11-3 a three-tube configuration, wherein each tube sees luminence video and one of the three primary colors. To accomplish this, a device called a beamsplitter is used. This can be a set of dichroic mirrors or, in some cameras, optical prisms. I will not try to explain the theory of dichroic mirror beamsplitters. I am sure that most of you have seen and handled a prism. Light headed for the blue gun goes straight through the prism to the blue tube. The red and green tubes see light refracted by the prism. (Each tube looks at luminence and one primary color.) Then the output of each tube is amplified and fed to the processing electronics. The output is a composite video/chroma signal. Why go to all this trouble? As we progress, you will learn the reason.

Section 4 (see Fig. 11-1) involves the camera viewfinder. In the HL-79D it's electronic. That is, it's a mini-monitor clamped to the side of the camera. By the way, in this configuration the camera is set up for handheld, shoulder-carried operation. The HL-79 and other professional cameras have option packages to convert them for studio use. One of these options is a different viewfinder, also electronic, that can be mounted atop the camera to its rear. This is called a *studio finder*, while the small one attached to the side of the camera is commonly called a *sports finder*.

Some home cameras also have side-mounted electronic viewfinders. In some cases these are integral to the camera design, while in other cases they are options. Other cameras have electronic viewfinders in the rear. There are some that have the look of optical viewfinders when in fact the viewfinder is electronic. Less expensive home cameras use through-the-lens optical finders, while bottom-of-the-line models usually the black-and-white cameras, use top- or side-mounted simple optical finders. I prefer the electronic viewfinder; especially the side-mounted sports finder type. The camera operator can rest the camera on his shoulder for added stability and ease of operation. An electronic finder even though it's black-and-white does permit the camera operator to view the scene exactly as it will appear during playback. It also, in many cases, permits the videophile to replay the tape into the finder immediately after recording for instant evaluation. Optical viewfinders cannot do this. Finally, if your camera has a zoom lens, either an electronic or through-the-lens, an optical finder is a necessity. A fixed or preset finder is only of value with a lens of the specific focal length for which the finder was designed. If your viewfinder does not "follow" the lens, you have no idea what you are actually photographing (what the camera is seeing). If you are shopping for a camera, the electronic view-finder is definitely worth the few extra dollars it costs.

Section 5 (Fig. 11-1) deals with filters. Most broadcast cameras have a minimum of three for various lighting conditions. Home cameras, on the other hand, usually have only one. In fact, some camera instruction books never mention filters at all, rather they talk in terms of an "Indoor-Outdoor" button or switch. When you change this switch, you are inserting or removing an optical filter via a mechanical linkage.

Section 6 (Fig. 11-1) deals with power consumption. Home video cameras are normally powered from the recorder itself (in portable configuration) or from an ac mains adaptor (used with an in-home unit). Very few use, or have provision for, external battery power. Professional cameras, on the other hand, are powered in one of three ways: battery power mounted on the camera or carried in a battery belly, ac power adaptors that slip onto a bracket in place of the battery pack, or an external source such as a CCU. We will not concern ourselves with either Section 7 or 9 as they are of little consequence to the video enthusiast, but Section 8 is.

Many cameras in the home video market do not permit lens interchangeability. That is, you buy a camera, it comes with a lens, and you always use it with that lens. Up until recently this was the rule, but some manufacturers like Sanyo and Panasonic are starting to break that rule by offering top-of-the-line home cameras with lens interchangeability. Usually the traditional type C mount is used, as in 16mm film photography. If you are planning to go further than just taping home movies, then a camera with lens interchangeability is definitely a worthwhile investment.

Speaking of lenses, I cannot see purchasing a camera that does not offer zoom capability. Whether or not the lens can be removed is not as important as being able to vary its focal length from a wide shot to a tight closeup. Going on a "shoot" without the ability to vary focal length is akin to using an old snapshot camera and trying to get magazine quality photos. We'll talk more about this in the chapter on making a home-video movie. If snapshots are what you want, fine. For a little more expense you can have one of the main features that the pro's have. The flexibility that comes with a zoom lens. By now you are probably beginning to see why I chose a top-of-the-line professional camera as our comparison standard.

Let's go on to the "Performance" column in Fig. 11-1.

HOW TO COMPARE HOME VIDEO CAMERAS

I'll start right out by saying that the performance figures for your home camera and that of the HL-79 will differ significantly. That's to be expected. At this point, we need a home video camera for comparison and one that really rates high in my book is the General Electric Model 1CVC2030E (shown in Fig. 11-4).

I like it for a lot of reasons. It looks a lot like the ones that the pro's use. In fact, that's part of the appeal of this camera and others like it. Let's face it, these days it's quite common to see a newsman in the street conducting an interview while a guy or gal with a shoulder-mounted camera videotapes the whole thing. As I stated earlier, mounting the camera's weight on your shoulder adds to picture stability. Sanyo also makes a series of cameras using this same basic design concept, and they are also very comfortable. Part of the RCA line uses this concept as well. This design places the viewfinder at your eye, where it belongs, while taking the weight of the camera out of your hands. This makes a lot of sense to me.

The General Electric camera has every feature you might want: a 6:1, F1.4, power-driven zoom lens; a choice of automatic or manual iris control; a fade-in-fade-out iris system; indoor-outdoor filter, and a lot more. It also features a camera-mounted microphone that telescopes away from the

Fig. 11-4. The GE home-video camera looks a lot like the ones that the pro's use and has good performance.

camera operator to avoid noise pickup from the camera or the operator. The microphone is highly directional toward the subject being photographed.

We ignored the weight and power consumption specs on the Ikegami a little while ago. Now let's look at them. Again, remember that the HL-79 and cameras like it are rarely powered from a VCR, while home video cameras usually are. Our broadcast camera draws about 30 watts (a healthy 2.5 amperes at 12 volts) while our home version by GE uses only 0.7 watt—quite a difference in power consumption. As to weight, the Ikegami weighs 6.7 kg excluding lens, while the GE weighs only 2.8 kg lens, microphone, and all. But there is a tradeoff.

The big tradeoff is in resolution. Remember that the Ikegami is a three-tube prism-beamsplitter camera. With optimum adjustment, this camera can produce over 600 lines resolution at center and in excess of 500 at the corners. While GE publishes no specifications on resolution, I was able to ascertain via the phone that it will do in excess of 250 lines center. For a single-tube camera in the home-video class that's very good. As another comparison, the Sony DXC-1640, a camera I have used in the past and one very popular in the educational/industrial field, claims only 300 lines of horizontal resolution. The Sony does have other features that justify its significantly higher price, but it's also a single-tube camera using a ⅔-inch Trinicon pickup tube of Sony's own design.

I say that a camera which will produce 250 lines or more of horizontal resolution is good. I base this on something I stated at the outset of this book. That while a TV station in the USA may be transmitting 525 lines of information, the viewer is lucky to see 250 to 300 of them. If, when connected through a VCR to a TV set you can get this figure, you will be more than content. Most of the better-quality home cameras meet this figure, which may make you wonder why TV stations don't use home video cameras for in-field work.

There are other reasons having to do with FCC-established broadcast standards, equipment versatility, and overall ruggedness. Simply, the home video camera could never withstand the kind of abuse that most professional equipment is subjected to. The very nature of the work dictates that professional equipment will be subjected to rugged treatment—recorders and cameras sometimes do get dropped. If this happens, the operator must be able to pick up the camera and continue working. He can only continue if his equipment has survived. A TV station pays for this rugged design concept as part of the price of day-to-day operation. Rarely are news spots lost because a recorder or camera fails.

Home video cameras for the most part require a lot more tender loving care. They were not built to withstand the day-to-day assault of electronic news-gathering work, nor should you expect them to. After all, $1000 or maybe even $2000 is a lot to pay for a home video camera, but it's not $50,000. Also, the commercial station needs the extra resolution of the multi-tube system. This is because raw field tapes are rarely aired as taken, rather they undergo editing and some degradation will take place. A field

Fig. 11-5. If you are used to a film-type home movie camera, then a unit such as the Panasonic PK-700A will make the transition from film to tape less painful.

cameraman may shoot a twenty-minute U-Matic cassette and find less than a minute of it makes air. To quote an old axiom: "That's show-biz."

I said earlier that I prefer a camera with a side-mounted electronic viewfinder. You may be quite used to shooting with a home movie camera, and there are many home videocameras that resemble their film counter-parts. The Panasonic PK-700A in Fig. 11-5 is one of those. It makes the transition for someone used to a Super 8 mm film camera quite easy. In fact, it very much resembles the latest Super 8 mm sound-on-film cameras. It also uses a single-vidicon pickup tube and will give close to 300 lines of horizontal resolution. Both it and the General Electric are full-color cameras.

Let's go back to our performance checklist for the HL-79D Ikegami camera (Fig. 11-1). We've already discussed resolution and most of the other designations really do not concern the home camera buff. I will, however, give a brief explanation of each: Frequency response is something that most people never think of in relation to a camera, but in dealing with electronic cameras we must. Like a stereo system, an electronic camera has a definite frequency response. In the case of the Ikegami HL-79D, it can "see" from 50 Hz up to 5.5 MHz. A top-line home video camera hits around 4

MHz, if properly aligned and properly maintained. Geometric distortion speaks in relation to the ability for a camera to accurately reproduce what it sees in geometric symmetry. No figures for this are published for home and low-end industrial cameras.

Registration is the ability of the three tubes to see the same point at the same instant in time, similar to *convergence* in a TV receiver. All three tubes must scan the same point at the same instant in time. If they don't, color ghosting will appear in the picture. To *register* a camera, a special chart is employed, called a RETMA Standard Resolution Chart. The camera is focused on the chart, which looks like a white checkerboard with black lines separating the boxes. Controls on the camera or CCU are then adjusted to eliminate color fringing. Registration is not a problem in single-tube home cameras. If what appears to be a registration problem occurs, then you probably have a bad pickup tube. Finishing the list are sensitivity and signal-to-noise ratio. The former relates to the amount of light needed by the camera to produce a usable picture, while the latter talks of the amount of noise (video, not audio) that will be present in the reproduced picture. Again, these specs are rarely published for consumer electronics, but the Ikegami specs give you a rough idea of what a professional camera will do. Even the industrial/educational Sony DXC-1640 mentioned earlier claims only 100 lux minimum illumination at F1.4 and a signal-to-noise ratio of 45 dB. Anything close to that is quite acceptable. In the next chapter I will deal with "Do's, Don'ts, and Questions" about video cameras.

Chapter 12

Video Camera Do's, Don'ts, Questions, and Answers

A DOZEN DO'S

1. Do keep your camera lens capped when not in use.

2. Do store your camera in a protective case when not in use or during transportation. Hard cases, well padded with foam contoured to the camera, are recommended.

3. Do allow your camera to warm up for at least 5 minutes before using it. In cold weather outdoors, allow an even greater time.

4. Do remember to set the indoor-outdoor filter to the proper position and white-balance the camera only after making the filter adjustment.

5. Do use a tripod whenever possible. This will add stability to your videotaped productions.

6. Do be sure you have at least the minimum lighting necessary to meet the specifications of your camera. If in doubt, consult your instruction book.

7. Do use an external microphone for your audio if you are any distance from your subject. Built-in microphones work well only when relatively close to your objective. Handheld and lapel microphones work well with adults; directional "shotgun" microphones are good with children.

8. Do keep your camera and lens clean and free from dirt and debris. If used in a saltwater environment, its advisable to wash the exterior of the camera, camera cable, and connectors with fresh water as soon as possible to prevent corrosion. Consult your instruction manual for the best procedure, it will vary from camera to camera.

9. Do be sure that all connectors are secure before beginning to tape. Double check before starting.

10. Do experiment. Remember, if you make a mistake, you can always rewind the tape and start over. Only time is lost.

11. Do review your tapes as soon as possible after making them. Be objective. Don't be afraid to recognize a mistake, note it, and correct it next time.

12. Do carry an extra charged battery when away from power mains. Murphy's Law states that your battery will die at the moment you want to use it. Be wise and beat Murphy to the punch.

A BAKER'S DOZEN DON'TS

1. Don't point your camera directly at the sun or any other bright light. You can destroy a pickup tube that way.

2. Don't drop your camera or hit it against any hard surface.

3. Don't operate your camera and recorder in a wet environment (such as rain) unless the equipment is rated for such service by the manufacturer.

4. Don't let children play with a video camera (or recorder). It's not a toy.

5. Don't *zoom* or *pan* very often or very fast. The results are annoying to watch.

6. Don't expect camera-mounted microphones to give as good audio as external microphones.

7. Don't keep a fully-charge Ni-Cad battery around, unused for months, because Ni-Cad cells will develop a memory. If in doubt, fully discharge the battery and re-charge it according to the manufacturer's instructions.

8. Don't open a video camera just to get a peek inside. There is nothing user-servicable in there, and you can damage a camera if you are not familiar with the proper service technique. If you must get a peek at the inner workings, write the manufacturer for a picture.

9. Don't try to fix a video camera yourself, (see 8 above) it takes special training and instrumentation. If you think there is a problem with your camera, take it to a competent technician.

10. Don't forget to store the camera, recorder, cables, and accessories after use. Leaving a video camera lying around the house is inviting disaster.

11. Don't loan out a video camera unless you know that the person borrowing it knows how to use it. Instruct the borrower on its operation if you have any doubt. Be sure to warn him about "Don't Number 1" above. Vidicons are expensive to replace.

12. Don't expect your first tape to look like it was produced by a Hollywood studio. Camera operating technique is something that must be learned. Learn from your mistakes, Don't get discouraged.

13. Don't forget to white-balance your camera before shooting. This sets the camera for the ambient lighting. Follow the directions that come with your camera; they are very important.

QUESTIONS AND ANSWERS ABOUT VIDEO CAMERAS

I have a very good home-movie camera. Can it be converted into a video camera? The answer is no. A video camera is totally electronic, while a motion-picture camera is a mechanical device. True, most modern motion-picture cameras offer certain electrical or electronic features such as automatic exposure control and a motor-driven film transport, but that's where the similarity ends.

Doesn't that make my home-movie equipment obsolete? Not really. While dollar for dollar (discounting initial investment) it is far less expensive to produce on tape than film, you might not want to invest the initial $1,000 or more in a video camera. For you it might be far better to continue using your home-movie equipment, and have the film transferred to tape by some outside company (especially if you have a lot of money invested in a high-quality motion-picture camera). There are times when it's advisable to shoot film rather than video. For example, there may be places where carrying a camera and recorder would be cumbersome. The single-unit construction of a home-movie camera makes it ideal for such occasions. In addition, any photographic techniques that you learn from making home movies can be applied directly to video photography. Your home-movie equipment is far from obsolete, even if you purchase a video camera, for another reason, which should be obvious to anyone who watches prime-time television. Take note of how many shows and made for TV specials are still produced on film. There is a special "look" to film that video cannot yet match. The two are separate and distinct techniques, yet they do cross paths in many places.

What can I use my home-movie equipment for? There are a number of avenues open to video photographers where film can be of immense value. One of these is in the creation of *special effects*. Currently, video special-effects equipment is out of the price range of most videophiles. In some cases, a home computer can be utilized for graphics such as *titling* and *supers*, but it's far less expensive to create such effects on film and then transfer them to tape. An excellent reference on this is *The Super 8 Handbook* by George D. Glenn and Charles B. Scholz (published by Howard W. Sams).

Why does a "professional" camera produce 600 to 800 lines of horizontal resolution, and a home camera only give 300, or so? The answer is simple: It's a combination of price and the differences that were outlined in the previous chapter. Your home camera has a single pickup tube. The professional electronic journalist usually has a camera with three tubes. This along with special electronic circuits and a superior-quality lens gives his camera a definite edge. I see no need for the average individual to spend the amount of money that it costs for a professional camera unless he plans on making the camera a source of income. The average viewer watching an average home TV receiver would never notice the difference. In fact, many viewers would probably say the picture coming from a home video camera is superior to what they see over the air. (Remember the discussion earlier in this book dealing with perceived picture resolution.) A simple conclusion is important: If the picture from your camera (or one you are thinking of buying) "looks good" to you, and the camera seems to fulfill your needs, then technical specifications mean little. You, and only you, can judge your needs, wants, and desires.

What's the best way to select a home video camera? The same criteria that hold in the purchase of any camera hold true here as well. First, it must suit your needs. If you are a casual videographer, then almost any camera will do. If you are a serious videophile or intend to use the camera for non-air production then your criteria will be far more strict. I prefer a camera that permits you to put the weight of the unit on your shoulder for extra stability when shooting (see Fig. 12-1). This is especially important when operating with the lens at full-telephoto position (called a *long lens*). The slightest movement of the camera is magnified many times in the finished tape (or film) and rigidity is very important.

Which is best, an electronic or optical viewfinder? I will review a few points. There are three basic types of viewfinders: *camera-mounted*

Fig. 12-1. Shoulder-supported camera makes for steadier pictures in playback (courtesy General Electric Co.).

89

optical, through-the-lens-optical, and *electronic*. The first will only be found on inexpensive cameras not equipped with zoom lenses. The other two are used on cameras that have variable focal-length lenses. For the casual videographer, a through-the-lens optical finder is sufficient, but nowadays, cameras equipped with electronic viewfinders are not that much more expensive. Furthermore, a camera with an electronic viewfinder offers the advantage of instant replay when used in conjunction with a portable recorder.

Are there cameras with built-in video recorders? Yes, there are, but they are not available in the consumer marketplace. Sony, NEC, and a number of other companies have developed integral camera-recorder units. Sony claims that these units will be on the market in a few years. The concept is that of using ¼-inch tape in a special mini-transport that is either an integral part of, or directly attached to the camera. The vidicon pickup tube is replaced by a solidstate device known as a *CCD sensor* to minimize weight and add simplicity. While demonstrations of such camera/recorder combinations have been given at some trade shows, it will be a while (possibly a long while) before you see them at your favorite discount house. They are still considered experimental.

Do I need a zoom lens on my camera? Unless you are a "snapshot photographer," I would have to say that a variable focal-length lens is a necessity. It gives you the freedom to create illusions not possible with a simple fixed focal-length lens. (Most cameras being purchased in today's market are so equipped.) Again, the same rules that apply to using a zoom lens in motion picture photography apply to video photography.

Some of my pictures seem to have light streaks in them. What are they and where do they come from? They are called *lag* or *image retention*. I will try to explain this highly-technical subject in a nontechnical way. To begin with, every video camera (regardless of its price) will exhibit some form of lag under certain lighting conditions. Usually lag occurs when a brightly lit subject has been shot against a dark background. The vidicon literally memorizes what it has been looking at. Then, when the camera is panned, or the subject moves, a mild form of streaking occurs. There is nothing wrong with your camera. This phenomenon is normal, and care with proper lighting technique will minimize or eliminate it completely. Consult your instruction manual, or if the subject isn't covered, write the manufacturer for suggestions. Most will be happy to help you.

What is the difference between Vidicon, Saticon, Neuvocon, Plumbicon, and other types of pickup tubes? For the most part, the differences are in two categories: First, the designations themselves are tradenames. For example, the designation Saticon is a registered trademark of Hitachi, Ltd. Likewise, "Plumbicon" is a trade name used by N.V. Phillips Corporation.

Just like automobiles, video pickup tubes comes in many varieties and the use of one or another does not necessarily make a camera better. There are many parameters to pickup tubes, and the state of the art is ever-changing. Some parameters that effect any pickup tube are the type of electronic-emitting design, the size of the target area, and the electrode configuration. Generally, the pickup tubes used in home video cameras have ½, ⅔, or ¾-inch target and are of tri-electrode design. A few, very expensive single-tube cameras can be found which use 1-inch target, but they are rare. Until recently, it was a rule in the video business to believe that a larger-target tube would yield better results, but modern design technology has produced ½-inch tri-electrode tubes that outperform their 1-inch counterparts of a few years ago.

Pickup tubes can be divided into three different groups for our purpose. At the top of the heap are *broadcast-quality* tubes such as those used in TV production cameras. The next step down would be *industrial-quality*, and finally consumer-quality tubes. Before you get upset at finding yourself at the bottom of the heap, let me hasten to add that the quality differences exist because tubes for differing applications require different design concepts. The pickup tubes used in home video cameras are just as rigidly engineered as those used in broadcasting. The main difference comes in what the tube is engineered to do. Keep in mind that the single-tube home-color camera was developed after the three-tube configuration for broadcasters had been well established. In fact, we have the U.S. space program to thank for its development. Remember those color pictures from U.S. manned space-craft? Remember how the quality of those pictures improved with each flight? The camera you own or are considering purchasing came into being as a direct result of NASA's need for a small, lightweight color camera for use on board spacecraft. That's not a bad percentage.

What's the best way to store and carry a video camera? The best and safest way is to purchase a hard-sided, reinforced, and padded case. The Anvil Company and a number of other companies make cases that are steel-reinforced and padded with heavy foam. Some have custom-cut interiors that hold a camera snugly; in others you must cut the padding to fit the camera.

What's the best way to connect the camera to the VCR? This will depend on the type of your camera and VCR. In the case of portables, your camera will probably plug right into a special ten-pin connector. The wiring configuration on this connector has been standardized among man-ufacturers for ease of equipment interchange. To be on the safe side, it's wise to consult the user manuals on both pieces of equipment. In most cases, it's simply a matter of plugging in and taking pictures.

The home video recorders (nonportable) are a different story. Most are not equipped with the ten-pin connector, rather, a set of jacks marked "video in" and "audio in" will be found somewhere on the machine. Older machines had them on the rear panel, while newer machines, to make them more accessible, have relocated the jacks to the front. To use most cameras with home machines requires that you also purchase an accessory "power pack". This unit adapts the camera to the recorder and supplies the power neces-sary for the camera to operate. Adaptor power packs are usually specifically designed for a particular camera, and it is not recommended that you interchange such units.

If your camera uses the older, nonstandard wiring configuration, serious damage to the camera could result from using the wrong supply. About a year ago, I had to fix a burnout in a friend's color camera after he tried to power it from a supply he had used with his old monochrome camera. It used the same plug, but the wiring was different. In fact, he was unaware that the high-voltage section of the old black-and-white camera was built into the supply adaptor and not integral to the camera. The result was a few burned out IC chips and transistors. He was lucky. He might have taken out the pickup tube. The cause? You guessed it—if you have the read book carefully thus far—the return video wire in the new camera was the same as the high-voltage wire from the old camera. The result was a burnout in the electronic viewfinder on his new toy. Most camera power adaptors are system-engineered for a particular camera. Use the right adaptor and the right power supply.

How far away from the recorder can I go with my camera by using extension cables? Not very far, unless you are willing to accept lower-quality pictures caused by cable losses and phase shifts that occur in long lines. In broadcasting, the cameras are designed to compensate for extended cable lengths. The latest technology permits a broadcast camera to operate up to a mile from its connection point into a video system through the use of encoded FM transmission via triaxial cable. Studio cameras usually have several hundred feet of cable attached to them. Studio cables are similar to that used on home cameras, only larger with more wires to control more camera functions. The video operator has, at his fingertips, controls that permit him to make adjustments for cable length, tune out unwanted phase errors, and *time* the camera into the overall video system. You have only a recorder and a camera. You don't have the special control system I have just described. Therefore, you are limited to the amount of multicore cable extension you can insert without significant degradation to your overall picture quality. It's suggested that you consult the camera manufacturer if you have any questions regarding maximum cable length. A good rule to follow is to stay under 50 feet. Another tip: Keep an eye on the red and magenta colors. They will be the first to be affected by excessive cable length.

What is a good support for my camera? A tripod designed to handle the weight class of your camera is best. Tripods are designed for specific maximum weight and, for example, you should never place a 6 or 7 lb video camera on a tripod designed to support a 2 lb, 35 mm camera. Even if the tripod will support that weight, the operation of the "Pan" and "Tilt" controls may be impaired. Many video camera manufacturers either sell or at least specify the best type of tripod to use. Follow their advice. Doing so will mean fewer problems for you and better pictures for your audience.

What's the best way to buy a video camera? This one is simple. First re-read the two preceding chapters of this book. Then, armed with the information you have gleaned, go to a store that sells many different kinds of video cameras. Work with a salesman until you find a camera that handles the way you like, performs the way you want and that feels comfortable. Remember, this is a big investment, so take your time making your decision.

Chapter 13

Your TV Set, the Last Link

I call this chapter *the last link* because it deals with the end product and the way you will view it. I'm talking about your TV set. We have already discussed some of the problems that can befall a neophyte because he is not aware that his TV set is in a state of disrepair or incompatible with his video equipment. It happens all too often.

In the hope of countering this kind of problem before you get caught in it, let me make some suggestions. As already stated, before buying a VCR, make sure your TV set can handle the machine and is in good working order. The same goes for your antenna, especially if you plan on doing a lot of off-air recording. Again I will state what I have said previously: If you have any doubts about your TV set, your antenna, or both, then call in a competent TV service technician. It's like taking out an insurance policy. I know this firsthand because I had to spend several hours performing modifications to some of my older sets a few years back in order to make them operate with a VCR. I was lucky, all it cost was time and parts. It could have been worse.

Let's assume that your TV set is either too old to warrant the necessary modification, or you just plain want a new set. Which way to turn? First, I am going to tell you that whatever you decide to buy, it will be solidstate. This means, except for the picture tube, all other active components in the set will be IC chips, transistors, and diodes. From the very first, solidstate TV designs have proven superior in reproduced picture quality to their vacuum-tube ancestors and appear to have greater longevity. This is because solidstate devices dissipate far less heat than vacuum tubes, and heat dissipation is one factor that determines the useful life of any electronic device. Also, in these days of penny-pinching, it's nice to know that solidstate TV sets use far less electric power than older sets.

There are two basic design concepts that have evolved in solidstate TV receiver technology. The first is a continuation of the single-unit-design chassis, although in solidstate sets this chassis is usually a printed-circuit board attached to a metal frame.

The other design concept also uses a printed-circuit board attached to a mainframe, but few components are on this board. It acts as an interconnect between smaller plug-in printed-circuit boards called *modules*. The main board is called a *motherboard*. There are a number of variations on this theme, including the use of a plastic mainframe and interconnect cables to eliminate the motherboard entirely. In both cases, tuners for vhf and uhf are externally mounted, as are some of the major power supply and deflection circuit components.

With the advent of computers and space-age technology, the very act of turning on and tuning a TV set to a desired channel has taken on a whole new meaning. Many of today's television receivers use microprocessor control for this and almost every other function. While turning the set on and controlling its volume level are not necessarily an important use of the microprocessor, being able to select a station without the use of any moving parts is definitely a step forward.

Most of today's television receivers, other than low-end models feature what is known as *varactor tuning*. The old-fashioned mechanical tuner package has been replaced by one that's totally electronic and has no moving parts. This means one less major headache because there are fewer mechanical parts to break down. A varactor tuner uses a special component called a varactor diode (see Fig. 13-1) in place of the usual mechanical detented-coil arrangement. A varactor diode is a tiny device with the property of varying interelectrode capacitance in relation to applied voltage. If such a

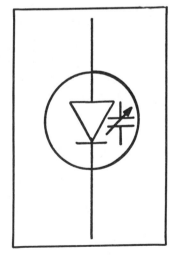

Fig. 13-1. Electronic symbol for a varactor diode.

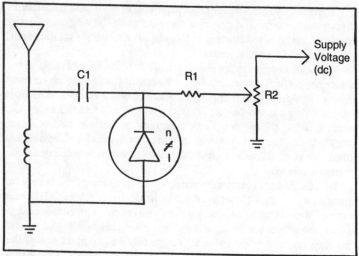

Fig. 13-2. Basic varactor-tuned circuit.

device is put across a coil (as shown in Fig. 13-2) then the resonant frequency of that circuit can be changed by changing the voltage applied across the diode. In this simple illustration, a potentiometer is connected between a dc voltage and ground, with the wiper used as the voltage-divider point. As we sweep the range of this control from zero to full supply voltage, the voltage across the varactor diode changes, in turn changing its junction capacitance. Resistor R1 is an rf-isolating device, while capacitor C1 acts to keep the dc voltage from being short-circuited to ground through the coil.

With this simple circuit, you have gone from many to one moving part, the potentiometer. But a potentiometer, unless it's assisted by some form of gear reduction drive is hard to adjust, especially when trying to tune a TV station. Enter the mating of the microprocessor controller to the varactor tuner (Fig. 13-3). Since TV stations are stable in regard to their operating frequency, then the designer can develop a *truth table* that states the tuning voltage requirement for a given television station. Now, if you could program a microprocessor to supply the exact tuning voltage for each given station, on user command, you would have a very precise tuning control. The desired channel number is selected by command of a touch-pad connected to the processor input. In essence, that's how direct-access digital varactor tuning works.

In reality, things are far more complex, but regardless of the electronic complexity (handled by one or two IC chips) the big advantage is the absolute minimum number of mechanical parts. This particular system also has another side benefit, that of *instant random access* to any channel from 2 through 83. This is because all tuning is done electronically, and takes place in the twinkling of an eye.

Another recent innovation to the consumer receiver is that of the *comb filter*. This neat device has been around video for some time. In fact, it predates a number of VCR designs that utilize it. A comb filter selects the best parts of a video signal, the parts that are wanted, and passes them on for further amplification, while rejecting the unwanted parts. The exact nature of this device and its operation are complex, but a comb filter in a TV receiver does add to the perceived resolution of the picture.

The little transistor has given way to the IC chip in many applications. The IC is an active device that combines many circuit components including many transistors into a single. The IC enables a significant reduction in the number of discrete components needed to make a TV set or VCR. Because of high-tolerance manufacturing, a more uniform design of TV receivers is possible. Again a benefit brought to you and me compliments of space-age technology. Solidstate technology has resulted in a higher-quality picture than ever before. Figure 13-4 shows one of today's "third generation" computer-age television sets, while Fig. 13-5 shows similar technology applied to the tuning system in a VCR.

While on the topic of television receivers, I cannot end without discussing yet another new trend, or I should say, the return of an old trend—that of large-screen projection TV. Large-screen TV using the projection technique is nothing new. In fact, RCA had a consumer monochrome set in a sample market in the late 1940s. It offered a larger picture than direct-view CRT sets, but its resolution when compared to the direct-view CRT left a lot to be desired. Over the years, projection systems were developed for broadcast and industry, but little attention was paid to the consumer marketplace until the mid '70s. Then the craze hit, sparked by General Electric and some other smaller companies. Various concepts were introduced and marketed, and soon people were buying or building TV sets that could give a wall-sized picture. The craze is only now starting to take root.

There are many possible configurations for projection television, but they all fall into one of two categories. Either they are rear-projection to a

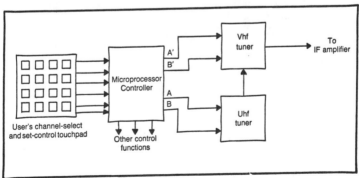

Fig. 13-3. Basic microprocessor-controlled varactor tuning system using touchpad user access. A and A' are bandswitch voltages. B and B' are tuning voltages.

Fig. 13-4. The ultimate TV? This 25-inch console television from General Electric is equipped with a 10-watt per channel stereo amplifier, stereo sources such as video disc and simulcasts. The set also features easy connection of peripheral video equipment such as video cassette recorders, video games, and auxiliary amplifiers. Equipped with Superband tuning, the set can receive up to 35 additional unscrambled, compatible cable channels without the need for a converter box. The set has infrared dual-mode remote control—even for cable channels. (Courtesy General Electric.)

translucent screen (such as the G.E. unit), or front-projection to a hyperbolic-curved reflective screen. Both systems work pretty well, but neither gives the overall quality of a direct-view tube—at least not yet.

The simplest and least expensive projectors marketed consist of a small portable TV mounted upside-down in an enclosure and a large lens in front of the picture tube (See Fig. 13-6). These units work, and you can build one yourself from kits or from scratch. The two most expensive components are the portable TV set and the screen. The drawback to this simple system is that the TV set was not designed for the task. Therefore, best results are obtained when viewing in a totally darkened room. Even then, the image produced may not be bright enough to suit your taste. It is however, a very economical approach to projection TV.

Most high-quality projection TV systems use three tubes that have been specifically designed for projection service. There is one tube for each primary color, and optical matrixing in the projection process combines the three images into a single full-color picture. Since three tubes are used, there will be some misalignment among the three images. This is handled by precise physical alignment of the three tubes (so that the images are superimposed, and then by an electron-beam bending process called *convergence* that eliminates any peripheral misalignment. The result is a full-color picture viewable in normal room light, but with a rather hefty price tag. High-quality three-tube projectors start around $3000 and go up from there. Projectors are not for everyone. Most projection systems look rather poor when viewed up close. They are designed for viewing at a considerable distance which means that you need a fairly large, dedicated viewing room. Nobody has come out with a standard of minimum viewing distance, but you can easily judge that for yourself by looking at a projection system operating in a dealer's viewing room that approximates the room you intend to use. Today's projectors are good and getting better with each model run, but it will still be a long time before they replace the direct-view picture tube.

Many improvements have come along that make a TV receiver of today a few hundred times better than its ancestor of ten years ago. Pictures are brighter, images are sharper, and best of all, modern technology has permitted manufacturers to keep the price of this item within the means of almost every person. Today's color receivers, with all the modern technology, sell for only a few percent more than a similar set a decade ago. Wouldn't it be nice if everything else were that way as well.

I cannot close without mentioning the type of receiver used at TV stations to look at the picture before it's sent out. Actually, these are not receivers, but rather monitors. A monitor differs from a receiver in that the tuners, IF amplifiers, and audio sections are deleted. Monitors concentrate on producing a very high-resolution picture and accurate color. If you walked into a TV studio you might be turned-off by what you see. The picture will probably not seem as bright as what you are accustomed to, nor will the color

Fig. 13-5. Varactor tuning and remote control in a VCR, Mitsubishi's HS-300V (courtesy Mitsubishi Electric Co.).

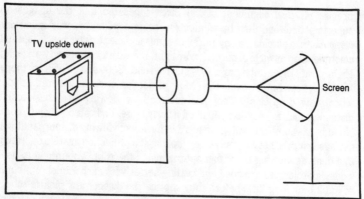

Fig. 13-6. Simple TV projector using small portable TV mounted upside down in front of a large lens and projecting onto a curved screen.

of the picture appear as saturated. The *hue* or *tint* of the color might also seem a bit off, but that's the way things are supposed to look. The picture that the engineer views is not set by eye. He utilizes some very sophisticated equipment (such as waveform monitors and vectorscopes) to be sure that the picture leaving the studio conforms precisely to the standards set forth by the FCC. The video monitor is calibrated as well, using a precise color-bar reference signal from a device known as an NTSC generator. It too has been pre-calibrated to a standard and is recalibrated at specific intervals. The picture on that monitor is exactly the way its supposed to look, and those monitors are far from inexpensive. Most 14- to 17-inch studio monitors cost far more than a 50-inch home TV projector. If you have no way of getting into a TV station to see things for yourself, you can adjust your home set to exhibit an approximation of the same. Here's how. Some night, after sign-off, tune around and see if any station is transmitting a pattern of color bars. If you happen across such a transmission, do the following:

First, turn off any automatic controls on the set. Then, fine-tune for the sharpest picture you can get. Next, adjust the color-level control to a point where all colors can be distinguished but none are saturated. Turn the contrast control to its approximate midpoint and the brightness control to something below what you would normally set. Now, here is the important part—carefully adjust the *tint* control so that the third color bar from the righthand edge of the screen is a pure magenta (purple). Do this very carefully, then turn the set off and go to bed. In the morning, turn it on and view the picture on the same channel. If the picture seems "dull" and "lifeless" don't be surprised. That's approximately the way most monitors look. They are supposed to look that way! To digress, as a result of having worked a good part of my life in video, I have become accustomed to that type of picture. To me, that's "good color," but I won't try to place my values on your televiewing habits. That's what your mind and the controls on your set are for. Televiewing is a purely subjective happening and very personalized to each individual. Nobody sees the same thing exactly the same way. That's the reason for the establishment of standards in the broadcast industry—subjective adjustment is not permitted.

Chapter 14
Video Software

Thus far, we have discussed only the machinery involved in home video—the hardware aspect. Now it's time to look into another very important aspect—video software. This discussion will take the view that there exist three basic forms of video software. The first two are apparent: blank unrecorded tape and pre-recorded entertainment. The third category is a not-so-new multimillion-dollar worldwide enterprise, that of pornographic home-entertainment software—the so-called "adult" films on tape. I consider "adult entertainment" to be in a category by itself, because it's not for everyone, and we won't discuss it here.

VIDEO CLUBS

Two aspects of home video that we will gloss over are the video clubs that seem to be springing up far and wide. In fact, let's get this discussion out of the way at the outset. Video clubs appear to be taking on two distinct formats. For sake of argument, let's give them titles of "Rental Entertainment" and "Experimenters" organizations.

When pre-recorded video software started to appear in the consumer marketplace, there was only one way to obtain it. That was to purchase it outright. Pre-recorded tapes of motion pictures and other events still can be purchased, but as many large corporations involved in this endeavor have noted, sales have never lived up to initial expectations. A pre-recorded videotape can cost anywhere from $35 to several hundred dollars. That's the legal sales end, not pirate copies of first-run movies still in general release. I hate to even broach the latter topic, but it's no secret that many films, especially in the science fiction category, have made many a pirate wealthy.

One need only read a newspaper to realize that the motion picture industry deems piracy a direct threat to the continued viability of the entertainment industry. As someone on that side of the fence, I wholeheartedly agree. I will state without reservation that rampant video piracy could easily spell the doom of the entertainment industry if left unchecked. I'll go into this in greater detail later on. For now, we are only concerned with the legitimate acquisition of video software.

Anyone who owns a video recorder/player is well aware that watching the same tape over and over again, even a first-run highly-acclaimed motion picture, will eventually get boring. Yes, it's nice to watch a film in the comfort of your own living room, but how often do you want to see the same flick? Many entrepreneurs were quick to realize that there would be a definite market for the rental of such software and established rental organizations. Initially, a cassette that retailed for a hundred dollars or more could be had on a 24-hour rental at about a tenth of that price. It does not take an expert to realize that ten rentals pays for the cassette and the rest is gravy. Many of the large motion-picture production houses recognized this, and formed subsidiaries to handle the wholesale aspect of video rental. Nowadays it is possible to rent tapes on any subject just about anywhere. Within a few years you probably will buy your groceries and rent your weekend entertainment at the same place.

Rental Entertainment clubs started as an offshoot of the retail rental business. In some cases hardware retailers anxious to sell more equipment established as an inducement, special "renter clubs." Sometimes the cost was born by the dealer in the hope of enticing a sale, and sometimes it was offered as a lower cost alternative to retail rental. Soon, many retailers realized that they had a goldmine on their hands. People who had purchased hardware elsewhere, or already owned video equipment, were anxious to take advantage of low-cost software rentals. The rest is history.

Most of today's rental entertainment clubs function in this way: First, you pay a set fee up front. This can vary from several dollars to several hundred dollars depending on the particular club, the type of services they offer, and the kind of software available. In some cases, the initial fee covers a given number of rental tapes, in other clubs it does not. From that point on, you are entitled to rent tapes at rental fees that are usually substantially less than direct retail rental. There are also special-interest clubs that handle tapes not available from any source other than outright purchase. If you are interested in this aspect of home video to any great extent, then a retail entertainment club can definitely save you money over the long run. By the same token, if you rent software only on an occasional basis, then you are probably better off staying with straight retail rental. The choice is yours. I do suggest that anyone considering joining such a club carefully explore all aspects of that organization before joining. Make sure that the services provided meet your needs before plunking down your hard-earned dollars. Again, be a wise consumer,

"Experimenters" clubs are just that. They are for people who want to use their equipment for video production and then exchange tapes with other videophiles. The organizational structure of these clubs is usually straight forward. In most cases such clubs are local, hold meetings where videophiles compare their latest productions, and are akin to other social groups. They may, or may not, have a dues structure, and often members pool equipment and talent toward the completion of a specific goal. This might be a video film for educational purposes, a series for public-access cable TV, or countless other possibilities. Some video clubs are actually offshoots of other organizations with many of them having their roots in computer clubs and amateur radio organizations. In terms of technology, the three are very closely allied.

Another aspect of experimenters clubs is the ability to get first-rate speakers who are well-versed on video production. It's very hard for an individual not in the video profession to gain access to such people, but many will freely spend an evening instructing and discussing their field before an audience. In essence, the video experimenters type of club is for the individual who wants to go a step beyond just recording his favorite TV program. It's for people who do want to gain an understanding of their equipment and its capabilities. These clubs probably won't make you into a top-rated $40,000-a-week Hollywood director, but the educational aspect they provide will be invaluable, if only to improve your skill as a home-video film maker. The technological and production aspects of video are changing daily, and a video club (along with some good video-oriented magazines) is a way to sharpen both understanding and personal skill. I highly recommend such organizations to those of you who want to take that "next step."

BLANK VIDEOTAPE

There are four basic grades of videotape available. Premium *broadcast mastering* tapes are the top of the line and are quite expensive. They are also not that easy to come by unless you happen to own your own video production facilities and purchase tape by the truckload. Nor are such tapes available in every size and length from every manufacturer. Such tape is usually not used by the videophile, since to the average consumer the cost cannot be justified. The next step down are *premium* "twins." Some manufacturers term this high grade while others call it "industrial quality." This is the *premium* grade for most consumers and costs only a few dollars per cassette more than regular grade tapes. *Standard* grade is next, and is the most popular. There is a fourth grade without any industry recognition, but, which can be purchased anyhow. Since it has no recognition designator, let's give it one: Garbage.

After this, you might be wondering just how to purchase blank tape cassettes. Here are some pointers. First, consult your VCR instruction manual and see what it says. The book will probably recommend that you stick to tape manufactured by the VCR-maker, and in most cases you can't go wrong heeding that advice. If nothing else, his tape will meet the specifications necessary for your VCR, but don't be afraid to experiment with other well-known, nationally advertised brands of tape. There is a key here that you must understand. A minimum guarantee of quality is insured on tapes which carry the VHS or Beta logo. These are trademarks, and those who own them have set specific standards to tape manufacturers in regard to reliability and overall quality. A manufacturer who cannot meet these specifications cannot legally use those trade logos on his blank tape cassettes. With nationally and world-known manufacturers, this logo will appear on all of their tape cassettes regardless of format or grade quality.

Tape I consider garbage is that which does not meet the minimum specifications as set forth by the licensor, and therefore cannot "legally" carry the aforementioned logos. Usually the tape in these cassettes is of dubious origin. That is it might be scrap videotape that's been bulk-erased and re-cut to fit the cassettes, possibly surplus computer tape which has undergone similar treatment, or tape from a variety of other sources. Cassettes themselves vary from exquisite copies of originals (but made of inferior material) down through cheap imitations that hardly fit into the VCR tape-loading compartment. What's tragic is that you are purchasing an unknown quantity when you buy unbranded tape such as this, and it can do harm to the recorder (especially the very delicate video heads). Tape technology is an important part of overall video technology, and reputable tape manufacturers know it. This is not to say that you will never get a "lemon" if you stay with a known brand. I've had my share, but a reputable manufacturer is one who adheres to good business practices by fulfilling his license agreement in regard to overall tape and cassette quality. "Garbage" cassettes usually are only a dollar or two cheaper than the known name, and for me, the saving isn't worth the potential aggravation. The fact is that many large retail chains often sell name-brand cassettes at a figure at, or below, wholesale cost to entice customers into a store in the hope of making sales of other merchandise. If you must pinch pennies when buying tape, keep an eye out for such sales in your local paper, and then purchase tape of known quality by the case. This is a way to substantially save on the cost of raw tape stock, and at the same time assure yourself of getting a quality product. Avoid "unmarked" or "unlabeled" tapes. They are headaches in disguise.

Obviously, the next question is what tape to use for a given situation. Let's start at the top. If you are planning to produce something on tape for general non-air distribution, and if it will undergo a number of generations from initial taping, through editing, and into distribution, then you will want to use broadcast mastering tape. Obtaining such tape may take a bit of research on your part, because it will not be available at your favorite discount store. In fact, you can only obtain broadcast mastering stock from distributors who service production houses, TV stations, and post-production facilities. Also, you may be required to make a minimum purchase, usually one or more cases of tape at a time. Few commercial distributors, if any, are willing to sell single cassettes in this category. The easiest format to obtain in this grade is ¾-inch U-Matic. It's available in time-lengths called *loads* in the industry, ranging from one minute to one hour. U-Matic format cassettes come in two physical sizes: smaller cassettes designed to fit portable equipment (maximum time 30 minutes per load) and standard-sized U-Matic cassettes to fit table-top VCRs (one hour maximum load length). Also, the smaller cassette shell will fit the table-top player affording one-way size compatibility.

In half-inch format, broadcast mastering cassettes are limited in availability, with few manufacturers willing to make the necessary expenditure for a produce line with currently limited sales potential. However, some Braodcast Mastering Beta and VHS cassettes are available in the Beta L-500, and in the VHS T-30 and T-60 lengths. They are designed to be used with equipment running at Betamax or VHS-SP speeds for optimum results. Run at slower speeds, improvement over high-grade tapes will not be apparent. This is a tape for the pro.

The best consumer grade tape, and one which will work quite well for the production-oriented videophile is that marked "HG" or "industrial." While not identical, they fall into virtually the same class and give more vivid colors and exhibit fewer drop-outs when compared to lower consumer tape graduations. If you are planning experimental production or are just looking for an improvement, even a subtle improvement, over lesser grades of tape, then this might be a way to go. At least it's worth the cost of a single cassette to find out.

Most of you will probably purchase standard quality tape for normal home use. This grade will give adequate performance on any properly maintained VCR, since as stated earlier, such tape must meet specified minimum performance standards before a licensor will permit the manufacturer to affix a VHS or Beta logo to it. For day-to-day home recording, this grade is fine.

There is no set rule for the lifespan of a tape. In broadcast news for example, where ¾-inch U-Matic is the base format, tapes are analyzed on a regular basis using sophisticated equipment to determine the number of damage spots and dropouts. When the number exceeds a pre-determined maximum, the tape is discarded. If a tape jams or the cassette shell is damaged it's discarded. Finally, if video heads seem to clog when a particular tape is inserted into a machine, that tape is thrown in the scrap heap. For the home video enthusiast who has no access to sophisticated tape, analysis equipment, the latter rules still hold true. If a tape exhibits severe dropout and gets worse with each pass, if the cassette shell or door is cracked or broken, or if a particular tape causes video head clogging, discard that tape. It's not worth the potential of an expensive VCR repair bill to conserve a few dollars and re-use a bad cassette.

PRE-RECORDED SOFTWARE

Pre-recorded tapes can be purchased or rented at retail outlets including VCR suppliers, camera stores, large chain stores, and houses that rent tape as this primary business. These days, you can rent or purchase just about anything, but the guidelines that hold true for blank tapes are also apropos in regard to pre-recorded tapes, especially when renting. A couple of moments spent visually inspecting a cassette can save you hours of grief. First make sure the cassette is the proper format. That is, it's VHS or Beta. Virtually all pre-recorded software will be either VHS-SP or Beta II so make sure your machine handles one of these formats. If the time-line of the recording load differs from this, it should be plainly marked on the package or wrapper. If you are purchasing a pre-recorded cassette, check to be sure that the package has never been opened. This might not always be possible, since some manufacturers purposely distribute in non-sealed containers. But, if the cassette is boxed and wrapped in clear cellophane, it's a safe bet that it's unused except during the initial transfer of material onto it.

If you are renting, don't expect a sealed box. In fact, you should open the container, examine the cassette for damage, and make sure that the tape inside is the one you want. Happenstance could lead to a cassette of one label being returned to an improper storage container. Once you are satisfied, be sure to read any literature that accompanies the cassette. In some cases, the renter will not want the cassette rewound. In fact, the industry is rumored to be developing a nonrewindable cassette for the purpose of limiting the number of times a rented tape can be shown per rental. Also note whether the tape has some sort of "electronic copywrite insurance," that is the tape has special signals that prevent it from being duplicated. These signals won't bother your playback VCR, but some older TV sets will roll or pull from them. If this occurs, try a tape that you have recorded yourself. If the TV seems normal, then tapes with these anti-piracy signals cannot be played back unless you have the set modified or purchase a new set which is impervious to them. Note: *Even tapes that are outright purchased may have some form of anti-piracy signal on them.*

Another thing to ask when renting a tape is who holds responsibility for tape damage. In most cases *you* will be the responsible party, and many rental agencies require a deposit in addition to the rental fee. The deposit is returned upon re-examination of the tape by the rental agency.

When purchasing tapes, look to see that the tape meets the minimum quality specifications as previously outlined. There is an exception here: Many large duplication houses use exceptionally high-quality tape, but while this tape is made by a major manufacturer, it may not bear that manufacturer's label. If it does, all the better, but read the "fine print" to be sure of the guarantee that comes with the tape. It will spell out the manufacturer's liabilities and your rights; the latter may vary from state to state in the USA and are different overseas. Also, deal with someone you trust, someone you know will listen if you must return a tape or exchange it. Again, this is common sense and it makes you a wise consumer. Finally, avoid pirate tapes since many of these are duplicated on inferior-grade cassettes. Think of it this way. If someone is willing to risk fines or imprisonment to "rip off" a producer of his profits, then why should he treat you any differently? Let's face it, a crook is a crook! Deal with righteous individuals and you will get your money's worth. Besides, many legal experts feel that purchasing a pirate tape makes you an accessory to a crime!

I should digress for a moment and explain that there are two methods used for tape duplication. They are known as *real-time* and *high speed*. Most home video cassettes for sale or rental still use the former, which accounts for part of the cost. Here is how the process works.

Regardless of the program source, a duplication master is struck on either 1-inch helical Type C or 2-inch Quadruplex format videotape. Most houses use 1-inch type C in the United States and Canada, while overseas another 1-inch format known as Type-B-BCN is used. No matter, a high-quality, enhanced transfer of material is made. This master reel is then played back on an appropriate machine into a processing amplifier and other equipment. If some sort of "anti-piracy" signal is to be added, it's inserted at this point. Then the video and audio signals are amplified by respective distribution amplifiers to a large number of Beta or VHS recorders. In large-scale installations, the *pause* circuits are tied to a common startup circuit. An operator manually loads tapes into each VCR, puts the machines into "record pause", and then starts up the master playback machine. He checks the monitor at each VCR to be sure all machines are receiving a signal and also checks the *set-up* of the master playback machine using a waveform monitor and vectorscope. This is usually done using color bars and audio tone recorded at the *head end* of the master tape. Once he is satisfied that everything is operating correctly, he sets his master tape to a pre-determined pre-roll startup cue point, remotely takes his rec-ord machines out of pause and into record, and rolls his master. Some houses duplicate hundreds of tapes at a clip this way. The system is called *real time* because it takes one hour to transfer an hour of material, two hours for a two hour presentation, and so on.

High-speed duplication is only now beginning to emerge on the scene, and few duplication houses can not yet afford the luxury of such devices for production of consumer tapes. A high-speed duplicator takes an edited master and physically presses it against a blank tape. The magnetic flux differential from this contact imparts a similar magnetic structure to the non-recorded tape. This system is used extensively for duplication of 2-inch Quadruplex tapes for syndicated broadcast and is only now gaining momen-tum in other formats. However, a number of manufacturers already have announced similar equipment for use with other broadcast formats and industrial use. Such machines for consumer tape duplication won't be far behind. Once this happens, a real price break will probably occur.

Chapter 15

Making a Video Movie

This chapter is for those of you who want to experiment with making your own home video productions. I will divide it into several parts. First I will deal with developing a concept with which to work and then following things through to the end product. In between I will discuss all aspects of video production on a professional level, and then try to apply them to home video production. This chapter, then, goes a bit beyond recording "Little Ginney's" fifth birthday party. It deals with the exacting standards required in video production regardless of tape format or equipment. To illustrate this, we will follow the production of a five-minute segment of a children's TV show—a show within a show that my partner, Burt Hicks, and I are putting together. I should add that producing a show does not guarantee it's going to ever see air. This is speculative production on our part, and all material to be discussed is copyrighted and therefore our property. It may never be seen on your local TV station. In fact, I am about to share a dream with you. Interested? Read on:

CONCEPT DEVELOPMENT

Note: The following conceptual material and the script material given in the Appendixes is copyright 1980, Bill Pasternak and Sanford Hicks, Pacific Educational Productions. Use of this material for video production is strictly forbidden under U.S. and International Corpyright Law. The material is presented for reference purposes only per authorization given TAB Books by the copyright holder.

Be it tape or film, there are many things that can be recorded for posterity. Either medium can be used for on-the-spot news gathering, covering epic events, recording history, education, or entertainment. Many times, tape or film shot for one purpose may wind up serving in several of these categories. However, each category is a science unto itself, and for each science there are specialists. For instance, some of the best video camera people I know are news-gathering electro-journalists. Yet, put them behind a studio camera, in a studio environment, and they cannot function as well as a studio cameraman trainee. Similarly, many of the finest studio-camera personnel are not good with a mini-cam on their shoulder. True, there are some who can get behind a camera anywhere and shoot tape or film in any environment, but nowadays the industry itself almost precludes such generalists, especially in big-city operations. Therefore, let us concentrate on one area, that of educational childrens' programming which entertains as well as enlightens.

A few years ago, late one summer evening I found myself pacing a line in my buddy's front yard. I had finished work on a film as its associate producer, and suddenly there was nothing to do but go back to a normal job; a rather mundane job, but one that put food on the table. I was looking for something to keep what pyschiatrists would term "the creative juices flowing," and was in a rather sorry state. It was 1979 and the middle of the energy "crunch."

One thing kept running through my mind. Every evening on the news there would be a story in relation to energy. Be it rising gasoline prices, projected petroleum shortages, or conservation, the news (local and national) was filled with energy-related stories. Yet, all of this was aimed at the over-21 set. Little or nothing was aimed at instilling the concept of energy conservation to the children of today who would be the adults of tomorrow and would inherit this problem. "Why not a childrens' show about energy conservation," I blurted out to Burt who was seated on the porch. That's how easy—or hard—depending on your point of view, it is to come up with a topic for production. This topic is known in our parlance as a *production concept*, or simply a *concept*. In this case, it would be designed to teach children the importance of energy conservation without making them fear the topic. In fact, it was immediately evident that if we undertook such a project it would have to be spoonfed in small, easy-to-swallow doses, and done in a way that the audience of children would not even realize they were being educated. Burt and I sound out different approaches and finally agreed that something akin to a five-minute segment of an already-established childrens' entertainment program might be the ideal vehicle into which a topical series such as this could simply be dropped-in. A short time later I was home behind my typewriter working on an initial outline.

THE OUTLINE

Once you have a concept firmly etched in your mind, there are several other preliminary steps. The first is to prepare a basic outline. Sounds easy? Far from it. In fact, it took well over two months before a workable thesis was complete. That exact outline is reproduced in Appendix A.

As you can see, writing an outline (especially one designed to obtain financial backing for a project) is not all that easy. It has to say, in as few words as possible, exactly what you are planning, why you are planning it, and how to accomplish this feat. Along with the basic outline, you must also develop a basic scenario. This is the working story itself. In the case of a single production it should tell the story completely. In our case, since we were planning to do 26 weeks of programming, it had to be expandable. That is, short and sweet, yet telling anyone who read it what the *story line* would be. That too was a part of my two month's work.

As you will note, it was decided from the outset that the shows would be produced using video rather than film. For our limited-budget operation, this choice was obvious since we had access to video equipment at reasonable rates. Burt was an experienced video cameraman and I knew enough about video editing to muddle through. Further, using video meant that we could complete our location shooting with a minimum of set-up time and personnel. The format chosen was 1-inch Type C with an alternative of ¾-inch U-Matic format if 1-inch became too expensive for this type of production work. We preferred the 1-inch format, then and still do. After a lot of deliberation, we elected to use "single-camera EFP" rather than hire a full crew. This would mean a lot more time in editing but still a substantial saving. The technique is the same as single-film-camera documentary or educational production.

Next came the task of developing a script, while at the same time deciding on the type of equipment, finding locations and talent, and developing a working script. Whether you are planning something along the lines of our production, or just a small mini-video film for your local civic group done on Beta or VHS, to this point and on through the scripting the approach remains the same. Where we do depart is in two categories: talent and equipment. Since we were planning, or at least hoping, for "air" for our work, we were forced to adhere to certain broadcast standards as outlined earlier in this book. Had this been a non-air presentation, we could have used less costly equipment and less experienced talent. We selected a young actress we considered to be ideal for the part. Older than the demographic group at which we were aiming, she had the ability to give total empathy to her audience. With the proper makeup we were sure that she could "pass" for the intended age we needed. An audition tape she supplied proved this beyond any doubt. With this out of the way, Burt and I turned to other matters.

EQUIPMENT SELECTION

As I said, we had to meet the rather demanding specifications of the programmers and of the FCC in regard to technical standards. By the time we would be ready to shoot the first piece of "air" tape for *You-Me and Energy*, we were assured that high-quality 1-inch portable equipment would be available. In fact, Sony Corporation had already announced its BVH-500 1-inch Type C portable recorder and a quick demo proved it to be the ideal piece of equipment for our needs. There was no way we could afford to purchase such a recorder, but rentals were available.

The "500" was selected to be our recorder of choice. Next we turned our eye toward a camera. Again, we would be renting. A lot of time was spent looking at various cameras and the quality we could expect from each. Also considered was weight, power consumption on battery power, and other aspects. In the end, it came down to one of three cameras, all in the multikilobuck range. After a final analysis, we elected to go with the Ikegami HL-79 camera equipped with diode-gyn Plumbicon tubes. (The runners up in my mind were the RCA TK-76 and Hitachi SK-90.) But cameras and recorders are only a part of what we would need. A list of equipment I knew would be imperative to carry on any *location shoot* is reproduced in Appendix B.

Did I forget anything? Yep, a whole raft of items that Burt added. In the meantime, we were trying to elicit financial support, but to no avail. As we added up our costs for rentals and outright purchases, we soon realized that without said support, there would be no way to tape anything. Yet, to get the support we needed, especially since we were an unknown quantity in the market vying against thousands of others with college degrees in TV production it became evident that we would have to prove our own worth. This meant having something to show a potential backer. At this point we should have been discouraged, but we were not.

SCRIPTING

Over the next ten months, we worked at developing and refining a couple of working scripts, looking at locations, and soliciting any help we could get. Scripting fell largely into my corner. That is, I would do the initial draft, give it to Burt for rewrite, then back to me for more rewrite, etc. We found that each of us working alone and conferring by phone could add more than *brainstorming sessions*. In script development, certain things are very important. Let's look at our parameters and constraints.

First we were talking with children. Therefore we had to avoid "big words" that they would not be able to understand. Any five-year-old can relate to a lamp, but explaining that by the use of terminology such as "thermoelectric conversion" was out of the question. The same held true for other words like "petrochemical fuels" and the like. Our verbage had to be easily understood.

Next came "delivery." Our feeling was that our narrator should be like an older sister or brother, not a parent or teacher. This would give a friendlier, more relaxed mood to the overall production, and would also make it easier for the audience to relate to the narrator. To do this meant "listening" to the way children converse with one another. I spent many hours at shopping centers eavesdropping on kids' conversations. This was done before one word was ever written. It's called background research on audience demographic reaction.

The final parameter was the way the script would "sound" with our chosen talent. For that reason, she was brought into the script development and her input was invaluable. At times a line that sounded "great" in my head

112

was lousy when read aloud. This meant rewording and inflection changes, a long tedious process. A sample of one of our final scripts is included in Appendix C.

I should digress for a moment and explain a bit about the format of the script. Especially the terminology being used. Its origin is film, and through the attrition of film people into video production, many of the terms have become common to both. Let's go through them.

¾-1 Shot means framing the picture so that you will see ¾ of the individual's body on-screen. In production, there are certain basic shots that are used. A *¼-2 Shot* indicates that the director is looking for a well-framed picture of two people in head and shoulders configuration. The first digit indicates the "size" of the shot, while the second digit spells out the number of people to be included. Hence a *½-1 Shot* indicates that a single person is to be framed from the waist up. Sounds simple? It is until you get into situations where one of the numbers can't be used. In this case, a director will "spell out" exactly what he wants. For example, he may want a facial closeup of a particular actor. Here he calls to his cameraman to give a "tight full face," or something like that. Each director has his own way of putting it.

The term *credits over* indicates that the title or other visual graphics information will be superimposed over the scene. This is done in *post-production editing* and not at the time the show is being recorded.

Mark could also be called location, a place where the talent must be at a specific time in relation to camera position. In the rehearsal process, the position of talent versus camera is tightly scrutinized so that both know where they should be in relation to one another at any particular instant. A *mark* is the position the talent must be in before starting to speak.

Sync should be self-explanatory. The meaning here is that sound and picture will follow lip-synchronization. *VO*, on the other hand, indicates that the particular scene's audio need not be in *sync* and will be *voiced over* at a later time. Again, this audio track is added in the editing process.

Tighten up and *pull back* should be self-explanatory. You might say "zoom in or zoom out." The meaning is the same. In fact, there is a whole glossary of terms used in video and film production, enough to fill an entire volume. These are just a few of the most common.

LOCATION AND TRIAL SHOOT

With several working scripts completed, we felt it was time to get something on tape. But a problem existed—one I mentioned earlier—that of finances. Commercial TV being as competitive a business as it is, there were no immediate backers forthcoming. Yet we were anxious to see what our "product" would look like. The cost of renting the equipment described earlier would have been over $2,000 per day, and that too was beyond our finances. After discussing the matter, we elected to go ahead and tape a trial run on VHS format using an industrial VHS portable recorder and a borrowed camera. The camera was a JVC KY-2000 tri-saticon EFP unit that sells for about $10,000. Had we not obtained this camera, we would have probably rented a single-tube home camera, since what we were going to tape would never reach air.

It was a make-do situation, but one of those times that bring out the ingenuity of human beings. For instance, rather than rent reflectors, we made our own. It was not hard, we simply took some heavy cardboard and taped aluminum foil to it. We made some reflectors with the dull side out for a diffused effect and others with the shiny side out for a hard-lit condition. As it turned out, both were needed.

Another make-do, or so we thought at the time, was in regard to talent. The young woman who was to act as narrator would not be available during the time she was needed, as she had other contractual obligations to fulfill. Almost at the last moment, we prevailed upon the thirteen-year-old daughter of another close friend to take the job. This young woman had no professional acting experience, but is an eloquent speaker with training in dance. We figured that with a bit of coaching she just might "pull it off." In the end, her lack of "total professionalism" proved to be an asset in that she was less structured than a professional might have been.

The script we decided to shoot was the *lead-in*. In fact, a version of the one in Appendix C, but rewritten somewhat for the style of delivery of our new talent. This meant we needed a location that would depict the scene as outlined. While California is noted for it's scenic coastline, many beaches are "off limits" for even amateur production. Others require special permits which vary in cost from municipality to municipality. It took a lot of searching by Burt and his wife, but we finally located the ideal spot.

How does one go about finding the "ideal" spot? There are many ways. In the case of many home-grown productions, your backyard, part of your home, or some public-access area might be ideal. For semi-professional work, you really have to give a lot of thought to the effect a given background will have on your overall finished product. For instance, our script and outline called for a "beach/ocean backdrop." But what kind of beach? One heavily congested with weekend sun-worshippers? Problems arise. First, the ambient noise level will be greater and it will effect the quality of your audio track. Is this the type of aural backdrop you might want? We felt this type of environment would be detrimental to the visual effect we were trying to achieve. Remember (a few chapters ago) my comments on audio and its effect on the overall video presentation. We opted for a more secluded location, one not overrun by tourists, yet easily accessible, and one that would not "break our pocketbook" in relation to permits.

The cost of obtaining a desired location is an important item to consider in a shooting budget. It's no secret that there are places that survive on what they make as "preferred locations" for all sorts of production work. Be it a private home, a store, private park—what have you. Before setting your sights in concrete it pays to look into every aspect of site availability, especially cost. As a case in point, a TV show I worked with for several months had an ideal location. It was a park, had exquisite scenery, and was free of charge. One proviso: it was free—as long as you did not damage any park property. To insure this, a deposit of $1,000 was required on entry to the premises, with the fee returned after inspection of the shooting location. Needless to say, all of us on that crew kept on our toes and never forfeited the deposit.

Even if you plan to shoot your "home epic" on your neighbor's front lawn, remember that you are a guest on that property. You also take on responsibility to insure that you cause no damage and if any problems arise, remember that you are the liable party. The following are some "do's and don'ts" for location shooting:

1. Do not run cables or place equipment where it can be a cause of injury to your crew, cast, any onlooker, or yourself.

2. Do bring only the minimal number of personnel necessary to complete the task. Sightseers and onlookers only get in the way.

3. Do try to rope off the area in which you will be working, to keep sightseers out.

4. Do be courteous but firm in dealing with sightseers and the "lookie-loos." Explain briefly what you are doing, but also tell them they must watch from the sidelines, off-camera, and keep silent.

5. Do be sure that you have some form of release form signed by everyone who appears in any video you shoot if there is any chance that the tape will be used for public showing EVEN ON AN AMATEUR LEVEL. This can and will avoid costly litigation at some future date. "Standard Release Forms" are available through most commercial video supply houses, or your attorney can draw up the necessary document. Don't do it yourself. It needs certain "legalese" that varies from state to state; country to country.

6. If operating on location from ac mains power, use only three-wire grounded cables. Never use two-wire cable or "two-wire to three-wire" adaptors. A good ground is essential for everyone's safety. If in doubt about mains power, go strictly battery power. If necessary, purchase, rent, or borrow spare battery packs for each piece of equipment requiring same.

7. Don't litter. You are a guest at any location even if you are paying for its use. Keep a supply of large plastic garbage bags with you and dispose of all of your trash after returning home. This is just common courtesy.

8. Do be sure that you bring every piece of equipment you may need with you to any shooting location. This includes such simple and often overlooked items as water and food.

9. If you do a lot of location shooting, do consult a reputible insurance agent in regard to personal and business liability insurance. A small investment in this can save you a lot of bucks later on.

10. Do scout a location thoroughly. Make sure any legal agreements needed are signed well before the schedules shoot date, that all permits needed are garnered, and that every conceivable obstacle has been cleared. Don't wait to the last minute and then find you have to cancel a shoot because you forgot a permit. Even if you are just a "hobbyist," be professional in your approach.

There is a lot more to location shooting than just these ten suggestions. I could go on at length, but instead I suggest that you consult any good book on professional film/video production, or if you are really interested in learning about this field, you might consider taking a course in the subject as an evening class. Many community colleges and other learning institutions offer such training at moderate cost.

Let's get back to the story. We have found our location, secured all necessary equipment, and the day arrives that we are going to do our tape. But exactly what are we going to put on tape. Again, refer back to the script (Appendix C). As you can see, we had only three scenes that had to be shot at this location. In fact, everything else would be existing 16 mm film footage, transferred to videotape and then *voiced over* on a separate audio track. How much work is there in shooting less than two minutes of videotape?

It took 7½ hours (not counting travel time). Why? Because it's rare that one is lucky enough to get what he wants the first time out. I am talking in terms of "planned production" and not "video home movies." Here is the sequence of events as they occurred in the 24-hour period surrounding the "shoot." We scheduled shooting for a Sunday after a Labor Day weekend. This would mean that there would be fewer kids and other onlookers at the intended location. On Saturday morning prior to the shoot, I assembled every piece of electronic equipment we would be using the next day. Each piece was independently checked out, and then the camera, recorder, microphone, and mixer were cabled together for a complete system checkout. Since we were using a three-tube professional camera, I performed a complete camera set-up using the specified procedure and registration chart. The recorder was disassembled from its case, thoroughly cleaned and reassembled. A static test of its operation was made using the manufacturers specified test-tape while viewing the pattern on a high-resolution color monitor with both waveform and vectorscopes in line. The microphones were checked, as was the mixer, and a second trial run was made. When all proved OK, the system was knocked down and packed into the proper carriers. Batteries were placed on charge.

While I was taking care of this, Burt was busily taking care of other important details: making sure that our "talent" knew where to meet us, checking his vehicle, procuring food and other needed items. When five o'clock Sunday morning came, all was in readiness.

"Wait a minute—5:00 A.M., you say? Why the rush? Simply because good production requires that adequate time be spent in setting up all equipment, setting scenes the way you want them, and anticipating a lot of other variables including "Mother Nature" herself.

At 6:00 A.M., all who were going on the shoot were assembled for breakfast at a predetermined location. In our case it was a coffee shop in Van Nuys, California. Over a leisurely breakfast, we discussed again every detail of what we were going to do and the responsibility of each individual.

An hour later, we were on the road headed north along the Pacific Coast Highway.

At 8:30 A.M. we arrived and set about finding suitable backdrops for our scenes. Several were available. We chose one that gave a backdrop of the Pacific Ocean, and, panning left, the beach itself. This would afford us use of the noon light, as we planned to shoot as much as possible with the sun directly overhead. While I and an assistant began setting up the equipment, Burt worked with the talent; rehearsing her again and again until he was sure that she could deliver the lines the way he wanted. A number of lines were rewritten on the spot. That's the reason for having a cardtable and typewriter along with us.

Equipment set-up followed this procedure: Everything was kept in cases until time for that particular piece of equipment to be installed. The tripod and head came first, next the camera was mounted and immediately covered with plastic wrap to protect it from sand, other debris, and salt spray. Next came the recorder and microphone mixer. They were cabled together, placed on the cardtable along with a small monitor, cabled together and covered with a clear plastic sheet. Next microphone and camera cables were put in place. The camera was cabled to the recorder; its battery pack installed and given a warmup period. My assistant set out a beach blanket, and "props" were placed thereon. These were simple—a radio, beach ball, can of "pop", and a towel. Exactly the kind of things you expect in a beach scene. Once all this was done, the camera was again powered-up, and after selection of the proper filter, it was white-balanced against the chart.

A microphone was then connected to the cable at the blanket, and a quick audio check was made. The microphone was fine, but we tried several different types until we found one that seemed to give the best audio quality with minimal wind noise pickup. The microphone chosen was a Shure Model SM-61 with a small foam rubber "Wind-Sock" over its head. The final preparatory step was to lay one minute of color bars or low-end industrial units. Here's why. One of the built-ins in a "professional" camera is a color-bar generator. It's controlled by a switch on the camera and is used to produce a video-level-reference signal on the tape. Its output level is preset to meet established SMPTE standards. The same goes for the audio tone, only it must be generated from another source. In this case, the source was a tone generator built into the mixer unit. The output level of our reference tone was set so as to produce a reading of "0 dB" on its output VU meter. Since the recorder had no metering, were were forced to assume that the levels generated by the camera and audio tone source were accurate (another reason for the prechecks the day before).

Our talent was put in place and her lines rehearsed. We set approximate audio level at this point. Once all of this had been done, we took a break. While "talent" romped in the surf, the rest of us huddled to keep from freezing. It was past 10:00 A.M., but the overcast persisted. You have not lived until you spend several hours in the clammy cold. The gal was smarter than the rest of us—the water was cold, but she actually was warmer once she became acclimated to it—score one for "talent."

We were *powered-down* until about 1:00 P.M. when the sun broke through, and the overcast disappeared. By this time, lunch had been taken care of, and it was time to make "TV." We powered-up the equipment once more, again white-balanced the camera and soon we were busily at work recording pictures and sound on tape. By 2:45 P.M. we were satisfied with our takes and by 3:15 P.M. all had been secured, the area policed for our trash, and we were headed home. But that's not where this all ends.

Once I arrived home, I immediately unpacked the camera, recorder, microphones, cables, and anything else that had been exposed to the salt air and carefully washed each piece of equipment until I was satisfied that any salt residue was totally removed. Salt is very corrosive and equipment exposed to such an environment should be thoroughly cleaned immediately after use. Everything was cleaned, including the interior surfaces of the recorder. Only after this had been done, and the camera returned to its owner, was the job completed.

EVALUATION AND POST-PRODUCTION EDITING

Later that evening, Burt and I got together at my house to view the tape. During the shoot itself, notes were kept on what we had recorded, and each scene had been *slated*. This means, that a piece of written information containing the scene number, take number, date, time, talent name, and director/cameraman names were recorded visibly on tape. Each scene had its own *slate*. We watched the tape several times, making notes as to which *takes* of a particular scene we deemed best. Once this was done, the project was shelved in favor of dinner.

Post-production editing on this project is not yet complete, but from experience I can take you through the steps. Since the tape was VHS format, there was no second audio track on which to lay *time code*. (This has already been explained.) We could go two routes: One would simply be to transfer our film segments to VHS in the order they were to appear. This would mean pre-editing of borrowed film for which we were responsible. Cutting someone else's film without permission could lead to problems. Choosing the other route, I elected to *bump-up* our master tape to ¾-inch U-Matic format. U-Matic has two audio channels, and this would permit adding time code on the second audio track. The film portions would receive the same treatment, thus permitting frame-accurate editing using any SMPTE time-code reader on almost any ¾-inch editing unit. Why didn't we shoot the original on ¾-inch and lay our time-code track initially? Remember, we had little in the way of funds and had to go with borrowed equipment for the most part. If a ¾-inch field recorder had been available, we would have grabbed it. At present, we stand with a VHS field master, an enhanced bump-up to ¾-inch time-coded, and a time-coded film transfer tape. Here is where any similarity between home production and "professional" production ends. When we have the funds to purchase the necessary editing time, the two ¾-inch tapes will be combined into a finished product using an editing technique commonly termed *by the numbers*. That is, we will select the exact numbered video frame for each scene to begin and end, select the effect we might want between scenes, select our credits which will be electronically generated, and program all this information into a computer. Also added on a separate, time-coded tape will be the rest of the voice-over dialogue, background music, and sound effects. A video editor operating this computerized editing system will put the final program together as we direct him to. No tape will be physically cut. All editing will be performed electronically, with the time-code permitting us to time every scene to a given frame—far more accurate than can be done in a home editing system.

But let's look at your situation. You have shot your tape. In fact, you have several cartridges of tape that comprise your home video production. How do you put it together? There are several ways, the best being to rent an industrial editing system for the format you are using. This type of system counts the control-track pulses recorded on the tape to enable synchronized playback. Each pulse is at a specific spot physically on the tape and corresponds to specific audio and video information. Using this system, you view your tape at normal speed, and make notes of scenes you want to keep and the approximate "numbers" on the display that correspond to the specific scenes. Keep this list, it's very important.

A complete basic editing system included a playback machine, a record machine, some form of *controller*, and the necessary interconnects. You may also have an audio mixer in the line. (See Fig. 15-1.) Using a system like this, two types of previously described editing techniques can be used. Either you can *assemble* scenes in order on a clean tape, or, you can *insert* material from one of your pre-recorded tapes on the other. I suspect that you will want to assemble-edit and keep your master tapes intact. In fact, its rare in high-budget that master tapes are ever directly edited. Usually high-quality duplicates are what the editor works with, while the masters are placed in a secure tape vault. You and I cannot afford that luxury, so to protect our masters, we will only assembly-edit.

The first step is to take a fresh tape, free of defects and record *black* over its entire length. There is a technical reason for doing this which is beyond the scope of this book; just assume it's what you must do as a first step. We are also going to assume that any titles and other graphic effects have already been recorded on your tape using standard photographic techniques (title cards, slide- or film-to-tape transfers, video from your home computer, or what have you).

Once you have your tape with *black* recorded, rewind it and begin editing. Remember you are using a "rented" industrial editing system and

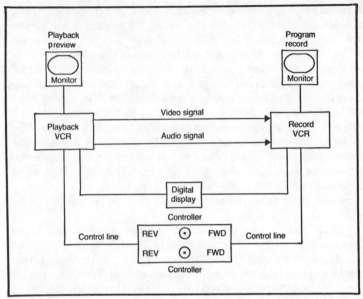

Fig. 15-1. Basic videotape editing system using control track pulse counting technique.

not two home recorders cabled together. I will explain the flaws in such a system later on. Let's assume that in Scene 1, the first video is number 235 and the scene ends at 467. Record that information. Scene 2 first video starts at 975 and runs to 1768. Advance the play machine to that number. How? The *controller* on the editing unit will have at least one, possibly two *joystick* controls very similar to those on home TV games. Here they control the forward-and-back motion of each machine. You advance your tape until you see the display reading the number you want. Now, if your rented editor has such facilities, you hit a button marked "Preview Edit." This will automatically roll the recorders back a few seconds in time, and then put both into play and simulate the edit on the screen of the "Program" monitor.

Let's suppose that you don't like what you see. You have not recorded or erased anything. Maybe you want to come into scene a bit earlier or later. Simply *joystick* to the new starting point on the play machine and again hit "Preview Edit." Once you are satisfied, hit the "Edit" button and Scene 2 will be recorded as you like, immediately following Scene 1. The rest of the procedure should be obvious. Simply keep going until you are satisfied with your work. When you reach the end of your final scene, roll back the edited tape to the beginning and view it. If you are satisfied, your job is done. I know it sounds easy, but don't expect it to work exactly as described the first time out. For most of you, editing tape even this simple way will be something entirely new. Like any art, it takes time to develop. Have patience with yourself and don't be afraid if you make a mistake. Remember, unlike film, with tape, you can erase and start again. Only time is lost. In this case, your time is an investment in education.

The question on your mind is where to gain access to even this simple editing system. The best place to look is in the Yellow Pages of your telephone directory under "Commercial Video Equipment Suppliers." Cost of a day's rental will vary with locality, equipment type, and accessories. At the outset, start with the basics.

Before closing this chapter, I want to touch on editing tape using unmodified home equipment. Yes, assembly-editing can be done to some extent simply by cabling two VCRs together, but don't expect "Hollywood" studio results. Home VCRs were not designed for this purpose. Some late models do have what are termed *transition editors* which simply back up the tape a bit when the unit is placed in pause, and then record over previously recorded material to gain servo-lock-up quickly. This is fine for putting together a home video log of family events, but will not do for high-quality video production.

The key difference between a home recorder and an industrial editing machine is in control circuitry, record lock-up, and the method used to erase previous material. An editing recorder usually uses something called a *flying erase head*. This is nothing more than a pair of erase heads that are mounted on the scanner assembly in such a position that they erase a line of material just prior to the record head inserting new material. In a home recorder and nonediting recorder/players of all standards, a full-track erase head is used which totally erases all material well before the tape ever reaches the record head. Remember my mentioning that the first step in using a full-blown editing system is to record *black* the length of the tape? Well, as far as we are concerned, a recorder that uses full track erase will simply erase our *black* and lead to dirty edits. By erasing on a line-for-line basis, as is the case in the editing system described, you will obtain clean, professional results. For the videophile experimenter, the rental of the equipment described and some good tutorial work can lead to excellent video-films at a lower cost than can be had using film itself.

As with everything else, video editing is as exacting as you want it to be. It can become a laborious job, or it can be fun. Either way, it will take time before you get the hang of it, so if you are planning on going the rental route, you might try to strike a deal for reduced rental rates by guaranteeing that you will be using the equipment on a regular basis. Many agencies welcome the chance to keep equipment in the field on slow weekends. Happy producing and editing!

Chapter 16

The Legal Aspects
of Home Video Recording

If you believe everything you hear these days, you would be afraid to live any sort of normal life. It seems as though people are trying to make laws to cover every aspect of life, home video being no exception. In fact, the legal battle surrounding personal use of a video recorder started a number of years ago, just after the first of the machines hit the consumer marketplace. Two giants of the entertainment industry, MCA Corporation and Walt Disney Productions, filed suit against Sony Corporation, charging that the sale of VCRs to the general public constituted infringement on copyrighted material.

What's this, those nice people who brought you *Cinderella* and *Pinochio* playing the heavy in this one? Not really, both MCA and Disney were only doing what any competent business would do. They were acting to protect their investment, in this case hundreds of millions of dollars in potentially lost revenues. So it was, that in the fall of 1981 the case reached the U.S. Court of Appeals for the 9th Circuit in San Francisco, where the three justices ruled in favor of the plaintiff. The court suggested that the manufacturers of home VCRs be compelled to pay a royalty fee on each unit sold as a method of offsetting potential loss of revenue to theatrical and other producers. The court also found that the use of a home VCR to record any copyrighted material was in violation of Federal copyright statutes, but it did not order that video hardware sales be stopped.

Where to, from here? Probably this case will wind up before the U.S. Supreme Court sometime in the future, and which way their decision might go is anyone's guess. In the meantime, the VCR industry is gearing itself for a long fight. According to an article that appeared in the December, 1981, issue of *Broadcast Management Magazine,* Sony's President and Chief

Operating Officer, Akio Morita, stated that his company would continue to market home VCRs and additionally, would continue to do everything possible to protect the public's right to use such devices. Similar words have been heard from other VCR manufacturers as well.

Meanwhile, on the legislative front, there has been some action to obtain relief for the general public. In October, 1982, a pair of bills which would amend the Federal Copyright Act to permit recording of copyrighted material for noncommercial use were introduced into Congress by Representatives John Duncan, of Tennessee, and Stan Paris, of Virginia. As this book is being prepared, the word is that the two bills are to be combined and sent on to committee for debate. It's known that the legislation is opposed by just about every aspect of the entertainment industry, including most major production houses, cable-casters, and satellite TV distributors.

While you might think this stand to be unfair, from the point of view of the distributor and those who make such vehicles in the first place, opposition to such legislation is definitely justified. The production cost of a "class A" motion picture is so high that any loss due to video piracy is intolerable. To spend 50 million dollars on a film these days is more the rule than the exception. Even the commercials on TV cost thousands of dollars to produce, and most of them run only 30 seconds. There is a lot of money at stake in the video entertainment world, and every special interest wants his cut of that pie—or more. What's tragic is that the consumer will be the one who must pay the cost for all of this in the end. For instance, should the Supreme Court eventually rule that VCR manufacturers must pay royalty fees to producers, directors, distributors, actors, etc., then these monies will have to come from somewhere. Obviously the manufacturers are not going to absorb this added cost. It will simply be passed on to the consumer as higher prices for video hardware and possibly video software as well. Nobody says this will happen, but it is a distinct possibility. There are others.

This is not the only battle looming on the video horizon. The premium telecasters are in a fight of their own with those they call *pirate decoder manufacturers*. As you are well aware, one of the hottest home entertainment items around is *premium subscription TV*. The vast majority of this programming is distributed via satellite to cable TV companies who encode it so that even the newest all-channel TV receivers can only see a scrambled picture with no audio. For a monthly fee, plus a nominal installation charge, the cable company will install a decoder box that permits you to view the programming. The other significant method of delivery is over the air, using the facilities of a vhf or uhf TV channel (usually uhf). Here the premium signal is scrambled prior to transmission and the premium telecaster provides the subscriber with a special antenna and decoder box to receive the pay-TV signal. National Subscription Television is the leader in this field with a number of other companies in the business as well. NST is also leading the fight against those who make a living by providing equipment to pirate the pay-TV signals, and who sell this equipment to the general public. The allegedly illegal units they are trying to have removed from the consumer marketplace are what the pay TV industry terms *pirate decoders*. These can best be described as "black box" decoders not authorized for use by the pay-TV company for interception of their signals, thus depriving the premium TV supplier of his rightful revenues.

The pirate decoder manufacturers, distributors, and even their users claim that once a radio signal of any type, even a subscription signal is transmitted it becomes *public domain,* and therefore if you have a technology to intercept and decode it, there is nothing that can be done to stop you from using that signal for your personal enjoyment. What legal maneuvering has occurred on this front? Quite a bit, but it has not yet even made a dent in the multimillion-dollar pay-TV piracy industry.

First to act was the State of California. In 1981 the state's criminal laws were amended to make it a felony for anyone to manufacture, distribute, sell, or even use any device that could intercept and decode pay-TV signals. Later in the year, Texas enacted a similar set of laws, with a number of other states soon doing likewise. The trouble is this: the Federal Communications Commission has been backing away from this problem. In fact, the only determination yet made by the FCC in this matter is that of *type acceptance*, an FCC term for putting their stamp of approval on certain types of electronic equipment before it can legally be used at both ends of any pay-TV system. This means that the FCC reserves the right of approval over both the transmission and reception equipment. Thus far, the FCC has yet to act forcefully against pay-TV pirates. Unless the FCC or some other Federal agency takes the initiative, I believe there is little the pay-TV industry can do to stop the pay-TV pirates.

The real "Catch 22" is this: If widespread piracy is permitted to continue, it could eventually cause large-scale failure in the new pay-TV industry. In the end, if the pay-TV companies succumb because they are unable to deal effectively with the pay-TV pirates then everyone will lose, including those who have purchased the so-called pirate decoders. But, the pay-TV industry seems intent on trying to solve this problem by legal means (enactment of laws to protect their special interest and amendment to the Communications Act itself). In my opinion, the same monies now being spent on large legal fees could be better used in the development of a high-technology pay-TV encoding scheme that would preclude the piracy problem itself. Possibly a single-chip IC decoder could be developed and guarded against piracy through proper corporate security and an effective encode/decode scheme. I happen to back the pay-TV industry in their fight against the pirates, but I believe that their methodology is definitely wrong. Currently, the only thing being accomplished is that of making a lot of "legal types" very rich.

Are pirate decoders legal? The bottom line seems to be this: The FCC says "No," but has done little else. The pay-TV industry says "No," and is trying to change laws nationwide in the hope of making this stick. The manufacturers and sellers of these devices say, "Yes," the devices are legal, and they intend to keep selling them until the Federal government forces them out of business, and the consumer couldn't care less. He sees the pirate decoder as a bargain that saves him the monthly subscription fee to the premium telecaster, not realizing that he is helping to destroy an important industry.

What about recording pay-TV signals? The laws that currently apply to the video recording of any copyrighted material apply here as well. While I doubt that the local constabulary will come knocking at your door to put you in irons for video taping your favorite movie, it is not inconceivable. If you elect to record pay-TV signals, know that you were warned of the possible consequences. Many pay-TV companies and cable-TV distributors will refuse to connect a home video recorder to their systems in a way so as to permit recording of premium TV material.

Chapter 17

Videodisc Systems

In late 1979, the Magnavox Corporation began test marketing a new home-entertainment device they called the Magnavision Video Disc Player. Since then, the Magnavision unit (which was a joint development project between Phillips Corporation and Magnavox) has been joined by two other entries in the videodisc race. The others are the RCA-developed system known as Selectavision and the VHD/AHD system developed by a conglomerate of Japanese electronic manufacturers. Each system has its good points and bad points, but from a consumer standpoint the "worst point of all" is that there is no compatibility of software between the three systems. The early days of home videotape were a nightmare of competition between many noncompatible tape formats, and videodisc seems to be following in that tradition.

The Phillips/Magnavox system is, perhaps, the most unique piece of electronics to come on the scene in modern times. It utilizes the properties of a coherent light source; a *laser beam*, in both the record and playback process. It is a totally optical system in which the playback assembly, or pickup is a beam of light. Except for the center-support hub, no mechanical contact is made with the videodisc itself, and the manufacturer claims that disc life is infinite (with proper care). (Pioneer also manufactures a unit, under license from Phillips, which will play the same videodiscs.)

The RCA Selectavision videodisc system differs in that it uses an actual stylus; similar to those used in regular record players, to track grooves cut in the RCA videodiscs. The grooves on a videodisc are different from those on an audio record, and the pickup itself does contact the disc causing disc wear. Also, the stylus requires occasional replacement, and this cannot be done by anyone not trained in the proper procedure. The big advantage of the RCA system is in the low price of the player and compatible software.

The third system is called *VHD/AHD*. The letters stand for video high density/audio high density, and this system combines some of the properties of both of the other systems. Like the Phillips unit, the VHD/AHD system is totally optical, but its pickup does contact the disc while playing back; hence there is some disc wear. The VHD/AHD system is only now entering its test-marketing stage.

Without getting into the technical aspects any deeper, each of these systems does reproduce a high-quality picture and good audio. The Phillips system offers certain advantages which we expect to see in later versions of VHD/AHD. Because the Phillips system scans rather than contacts the disc, it's possible to include such features as *freeze frame*, *slow-speed crawl*, and other goodies that would damage a disc tracked by a stylus. The Phillips system also includes two separate audio tracks with frequency response equal to the finest high-fidelity stereo equipment available today. There are technical reasons for this, but they are beyond the scope of this chapter.

The biggest hang-up in the proliferation of these home entertainment units is not the machines themselves, but rather the current state of available software. While pre-recorded software is of high quality for both the RCA and Phillips units, the overall selection is not yet large enough to satisfy equipment manufacturers. Unlike videotapes, it is not easy for a company to simply set up shop and start producing videodiscs. The manufacturing process requires a multimillion-dollar capital investment in tape and disc equipment; and the needed equipment is limited in source availability.

To date, the hoped-for "boom" in the sale of videodisc systems has yet to develop. Statistics released by a number of videodisc manufacturers for 1981 were very disappointing, and dealers that I spoke with in preparation of this chapter told me that there seems to be consumer apathy toward videodisc. It was pointed out that in any of today's videodisc systems, the maximum *per-side* playing time of any disc is 60 minutes, then the consumer must turn the disc over or change discs. Consumers simply don't seem to want the annoyance. They want a "push the button and walk away" playback system that can handle a two-, three-, or even six-hour program, just like tape.

Other consumers concede that the visual quality of videodisc is superior to ½-inch tape, but cite the inability to do their own recording on videodisc as a reason for choosing tape. Finally, you need only look at all the ads in the newspaper to realize that while "retail list" pricing on videodisc players and software is lower than that of tape; the many mass-discounters have changed this to some extent by offering cut-rate pricing on low-end tape and VCRs. I have purchased 2-4-6-hour "name brand" VHS blank cassettes for under $10 in single-lot quantities from a number of such places during advertised sales. There is almost no differential in actual consumer cost between a "low-end" stripped-down VCR and a videodisc player. At least on a retail level, the highly acclaimed cost advantage of videodisc over tape has been somewhat negated, to the detriment of both industries.

Is videodisc the "record player of the future"? Three years ago, everyone was predicting it would be, but these days nobody is willing to stick their neck out that far. The final answer will not be known for several years, but the hardware and software already in the consumer marketplace is good and improves with each model run. If videodisc can overcome the current consumer apathy toward it, the future for that industry will be bright, but the industry will have to listen to the public and give them what they want.

Another avenue for videodisc, now in the experimental stage is helping the consumer to shop via videodisc catalog. Sears has already taken the initiative with a pilot program to place its entire catalog on videodisc for use in catalog sales centers. The system chosen was the Phillips laser recording system; which is ironic because Sears markets a different type of videodisc player.

With that you have the world of consumer video as it stands today, except for the topic of home satellite receivers. The latter could fill a text in itself. I hope you enjoyed this book, and above all, I hope you learned a bit from it. Your comments are welcome. (You can reach me by writing to me in care of The Westlink Radio Network, 7046 Hollywood Boulevard, Hollywood, California 90028. Please include a self-addressed, stamped envelope for a reply.)

Appendixes
Video Movie Documentation
for "You-Me and Energy"

Appendix A
Concept

You-Me and Energy is a series of 5-minute educational "drop-in" segments designed to be tied to existing of planned Saturday A.M. children's programming. The material is educational in nature and is designed to educate the viewing audience to the problems of the current "short range" energy problem, as well as the long range energy crunch.

Our approach is one of "being positive about the future," that we as a nation can solve this problem, and that it will probably be one or more of those viewing the program who will arrive at the needed solutions when they reach adult maturity.

The program is demographically aimed at a "5-to-12-year-old, male/female audience." It will be hosted by one or more on-screen narrators, who will appear to be just a bit older and wiser than the audience itself. All narrators/host will be female. Narrators simulated age group will be between 15 to 18 years of age. Professional talent will be utilized.

It is well known that to hold a child's interest in relation to an educational topic such as this, the program must also be entertaining. To that end, the producers feel a combination of live action combined with an easily understood graphics can hold the child's interest for the span of the program and impart the necessary educational value. It is based upon the foregoing, that the producers feel videotape rather than film is the proper medium for such a presentation. This because videotape is totally compatible with electronically generated graphics and provides superior quality than can be obtained on film (16 mm). The plan is to produce this program on either ¾-inch or 1-inch helical scan videotape, permitting relatively low cost post-production and an overall high quality product which is also cost effective.

SCENERIO

You-Me and Energy will be divided into 26 five-minute bites. It will begin (show number 1) with an explanation of the word "energy" itself and continue (shows numbers 2 through 26) by detailing sources of energy, the conversion of energy sources into work, exploration for new, alternate, and renewable energy sources and the need for self motivated energy conservation. The program will always accent the positive concept of "yes we can solve the problem," and throughout it's structure it will stress the point that no problem is too big to be dealt with.

A secondary though related point covered will be that of human energy, and thereby physical fitness. In most segments (though not all) our narrator will refer to how "we as humans expend human energy" and then tie such to the dialogue of the particular program segment. We will then enter the discussion phase of the particular segment which will utilize footage descriptive to our topic, or graphic presentations related there to. The latter will use computerized animation. To close, we will return to our narrator at the same setting as the opening and relate our closing to both the primary and secondary objectives of the program. Opening and closing dialogue will be sync. All other dialogue is "Voice Over." Editing will be electronic.

SUGGESTED TOPICS TO BE COVERED:

1. What is energy? What does energy mean?
2. Man's discovery and initial applications of energy
3. Human energy—the human body
4. The energy revolution (as related to the industrial revolution)
5. Where energy comes from (overview of sources of raw energy)
6. Oil and petrochemical fuels
7. Coal and other fossil fuels
8. Electricity
9. Alternate energy: solar power
10. Alternate energy: atomic power, pro and con
11. Alternate energy: geothermal power
12. Alternate energy: the wind
13. Alternate energy: the sea
14. How man wastes energy, and why
15. Life without energy
16. Overview of how we use energy
17. Energy as a source of heat and light
18. Industry and energy
19. What is an oil crisis
20. Why America is oil dependent
21. How America is breaking its oil dependence
22. Energy conservation—what is it?
23. What the citizenary and government can do to become energy independent

24. What the child can do to help the nation become energy independent
25. Child self-motivation toward energy conservation
26. Our future if you solve the problem

NOTE: The concept of this program is education. Based upon this, no attempt will be made to veer away from subject material deemed controversial such as the atomic power controversy. While input will be solicited from interested parties, content of the program will be at the discretion of the producers and not subject to influence by any and/or all special interest or lobby groups. The autonimity and therefore the value of the program as an educational tool is far more important than the beliefs or views of any single individual or pressure group. The program will make no value judgments, but instead present data as known and accepted by competent scientific and governmental authorities, educators, and other accredited sources, Such material shall be written into language understandable by the audience demographic age group and thus presented.

Appendix B
Equipment List

Following is a list of equipment I deem necessary for location shoots for *You-Me and Energy*. It covers for every eventuality.

Camera. Ikegami HL-79 with 3 spare battery packs and rapid charger and ac power supply. Lens is 14-1 Cannon or Fujicon with 2X extender.

Recorder. Sony BVH-500 with 3 sets BP-90 batteries and charger/ac power supply.

Tape. Scotch-3M 1-inch Type C Broadcast Mastering.

Tripod. Any high quality unit with attached dolly and Miller fluid head.

Microphones. Shure SM-61 (X2) hand; Sony ECM-50P (X2) lav; Senheisser 815 Shotgun with fishpole and wind sock. Plus battery packs and batteries for the Senheisser & Sony microphones.

Mic Mixer. Shure M-67 with battery supply and batteries.

Cables. Camera- 50' min. (10-foot lengths with connectors). Microphone: 2 = 50' with XLR Male/Female.

Audio Recorder. Sony TC-126 with supply of Scotch, 3M AVM series C60's and patch cables to mixer.

Lighting. Kool-Lite set for indoor shoots; with 50-foot extension (3 wire grounded) for each lamp. Also 6 spare lamps. Reflectors for outdoor fill.

Miscellaneous

Card table	Typing paper
Sand bags	Make-up kit
Gaffers tape	Towels
Typewriter	Food, coffee and soft drinks

Chalk

Ac junction boxes (3 wire)

Garbage bags with ties

Earphones (2 sets)

Color field monitor, 5-8 inch with battery and ac supplies

Cables for above

Appendix C

Script

Scene 1. Music up on black. Fade to wide angle of deserted beach. Lone figure emerges from water and runs toward mark. Camera tightens to ¾ one-shot. During this sequence, credits are over. Music down on cue, of Heidi reaching off camera for microphone.

> Heidi (sync): Now that's the kind of work I like. Swimming makes me feel good and helps keep me in shape. It also uses up a lot of energy. Human energy in this case.

(At this point Heidi changes position. Shot adjusts to new position and maintains ¾ shot and comes to tighten as we progress.)

> Heidi (sync): You know, we hear the word energy a lot these days. Every time you turn on your radio or television set, there's always someone talking about energy. But, do you really know what the word energy means?

Scene 2. Cutaway to open dictionary. Finger is seen running down page until it reaches the word energy, then holds. Camera tightens.

> Heidi (VO): Well, the dictionary tells us that energy is vigor of action, or effective power and the ability to do work. Sounds a little complex I know, but maybe we can simplify things a bit.

Scene 3. Single auto on street. Follow auto onto freeway, or crowded street.

> Heidi (VO): Take this car for example. It's one way of getting

136

from place to place. When it moves along the road or street, it's doing work. It's work is moving you from...oh...let's say your house to your school. To do this, it's taking something called gasoline and actually burning it in a special part of the car called an engine. You might say that the gasoline is the place where energy is stored, and the engine is where it is changed into the power to make the car move. This fits the definition of energy because the effective power stored in the gasoline is making the car do work.

Scene 4. Any electrically operated machine: several shots in sequence.

Heidi (VO): The same is true of this machine. It changes another form of energy into work. In this case the energy is called electricity.

Scene 5. Lamp with shade removed. Hand moves into frame and turns lamp on.

Heidi (VO): The most common way we use energy is in lighting our homes. Each time you turn on a light, you change one type of energy into another. In this case, electricity into light.

Scene 6. Photo of Einstein with various scientific formulas.

Heidi (VO): In fact, a famous scientist named Albert Einstein proved two things. First, that in the entire universe there is a definite amount of energy. And...second, that human beings like us can't make or destroy energy. Yes, we can change it's form and in doing so make it do work for us, but that's all.

Scene 7. Rush hour traffic on freeway.

Heidi (VO): If Mr. Einstein was right, then why all the current fuss about using as little energy as possible? All this talk of energy conservation. Shouldn't there be enough energy to last forever? He did say that we can't make new energy or destroy old energy didn't he? Why then all this fuss?

Scene 8. Auto getting gas. Several views intercut.

Heidi (VO): It's simply that once we burn this gasoline in a car engine to convert it into motion, we can't use it again. Burning gasoline makes a lot of heat, just like any other fire, but in this case the heat is wasted. That's why we have to have the gas tank refilled every so often. We can only use the gasoline once. Make it work once.

Scene 9. Static photo of oil well.

    <u>Heidi (VO):</u> And...there is only so much oil in the world from which gasoline is made. The same holds true of most other energy sources...except...oh...the sun and the wind. But, once we use up all the oil in the world, there won't be anymore gasoline. It will have been changed into things that can't power our cars, trucks, ships, and airplanes.

Scene 10. Heidi again on beach. Standing with mic in hand.

    <u>Heidi (Sync):</u> The meaning then of "energy conservation" is to use what we have wisely, so that it will last as long as possible. In future episodes, we will show you some of the ways energy is made into forms we can use to do work for us. Also how to use it in the most efficient way. Right now I think I'm going to use a bit more of my own energy and take another swim. See y'a.

Scene 11. Heidi hands mic off screen, turns and runs toward water. Camera pulls back, music up, credits supered, end credits, fade to black, music down and out.

Index

A

American Radio Relay League, 73
Ampex Corporation, vii-ix, 5, 10
Ampex 2000, 19
Ampex VPR-20, viii
Amplifier, audio, 14
Antenna, 53, 56, 94
Antenna installations, 30
Aspect ratio of film, 69
Audio amplifier, external, 47
Audio dub button, 27-28
Audio mixer, multichannel, 66
Audio quality, enhancing, 66
Audio quality, poor, 67
Audio tracks, multiple, 28
Audio treatment, 47
Azimuth helical recording, 37

B

Band splitter, 30
Bandwidth, 5
BBC (England), 10
Beam-splitter, 61
Beta format, 1, 20, 22, 58
Betamax recorder, 1
Betamax II Zero Guardband format, 36
Beta speeds, 22
Beta tape handling system, 32
"Black boxes", 58
Blocking, 78
Brainstorming, 112

Broadcast equipment, ruggedness of, 83
Broadcasting, vii

C

Cable TV, 55, 74, 123
Cable TV premium channels, 72
Cable TV, taping from, 72
Camera, purchasing a video, 92
Camera, selecting a home video, 89
Camera, storing the, 91
Camera, three-tube, 60
Camera and VCR, connecting, 91
Camera comparison, 82
Camera control unit, 78
Camera don'ts, video, 87
Camera do's, video, 86
Cameras, home versus industrial, 60
Camera support, 92
Cameras, video, 60, 75
Cameras with built-in video recorders, 90
Cartravision, 23
Cassette tape, 20
CB radio interference, 72
CCE sensor, 90
Chiron device, 2
Chroma, 4
Circuit use, 43
Cleaning tapes, 57
Clock, internal digital, 31
Closed captioning, 8

Clubs, video, 102
Color ghosting, 85
Color signal, 44
Color slides on tape, transferring, 69
Comb filter, 64, 96
"Commercial killer," 62
Composite signal, 78
Computer interference, personal, 73
Controller, 120
Convergence, 99
Copy Guard, 58
Copyright laws, 72, 122
Copyrighted material, recording of, 123
Cost of video equipment, 7, 29, 75, 83
Credits over, 113
Crosstalk, 37
CTR TV, 97

D

DBS, 74
DBX audio-enhancement system, 47, 66
Dc restoration, 62
Detector sensors, 32
Digital counter, 28
Disc recorders, 126
Distortion, geometric, 85
Dolby, Ray, 11
Dolby audio-enhancement system, 47, 66
Dropout compensator, 45

E

Earnings, 63
Editing, viii, 65-66, 113, 118-119
Editing, field frame accurate, 28
Editing, post-production, 113, 118
Editing equipment, 66, 119
Editing equipment, home versus industrial, 121
EIAJ-1 international agreement, 20
Eject button, 27, 29, 32
Electricity, ac, 15
Electromagnets, 15
Electronics Industry Association lobby, 73
Equipment costs, 7, 29, 83
Equipment for making movies, 111, 134
Equipment set-up for filming, 117
Experimenter organizations 102-104

F

Fast forward button, 27, 29

Fast scan, 50
Fast scan (FSTV), 4
Fast speed, 18
FCC, 124
FCC broadcast standards, 83
Film chain, 69
Filming distance, 92
Filters, 81
Format, choosing a, 50
Format comparison, VHS and Beta, 32-35
Format lifespan, 58
Freeze frame, 9
Frequency response, 84

G

GE 1CVC2030E video camera, 82
GE 1VCR2002X recorder, 29
Generator, electric, 16
Ginsburg, Charles P., 10
Goodspeed, Rupert, 6
Graphic equalizer, 66
Guard band, 37

H

Head azimuth, 38
Head gap, 36
Head height, 38
Heads, dirty, 57
Head size, 37
Head system, four-head, 61-62
Head system, two-head, 61
Helical scan, 7, 19
Heterodyne converter, 45
Heterodyne video recording, 44
High-pass filter, 72
Hitachi, 79
Hitachi, SK-90 video camera, 112
Home movie cameras, converting to video camera, 88
Home-movie equipment, 88
Home movies to tape, transferring, 68
Hue, color, 100
Hum, noticeable, 68

I

Ikegami HL-79 video camera, 112
Ikegami HL-79D video camera, 75
Ikegami video camera, 83
Image retention, 90
Industry standards, 20, 52, 73
Input signal, 75
Instant random access, 96
Instant replay, vii
Integrated circuits, 20
Intercome, 79

140

Interference, 72-73
International agreements, 20

J

Jacks, 29
Japanese Victor Corporation (JVC), 61
JVC, 61
JVC KY-2000 video camera, 113
JVC 6300 recorder, 21

L

Lag, 90
Laser beam, 126
LEDs, 35
Legislation 73-74, 122
Lens, zoom, 81, 90
Lens interchangeability, 81
Loading stress, 33
Loads, 105
Location for shooting, obtaining, 114
Location shooting, 113
Location shooting do's and don't, 115

M

MacDonald, Copthorne, 4
Magnavox, 47
Magnavox Corporation, 126
Maintenance, 51
Maintenance schedule, 58
Mark, 113
MCA Corporation, 122
Microphone, 14
Microphone, use of, 117
Microphone, wireless, 68
Microphone rental, 68
Microphones, 67-69
Microprocessor controller, 43
Mitsubishi, 9, 47
Mobile units, broadcasting, 6
Monitors, 99
Morita, Akio, 123
Motherboard, 95
Motion, low, 9
Movie concept, creating a, 109, 131
Movie outline, writing a, 110, 132
Movies, making video, 109
M-wrap, 33

N

National Subscription Television, 123
National Television Systems Committee (NTSC), 4
Neuvocon, 90
Noise reduction system, 66
NTSC, 4

NTSC generator, 100

P

PAL system, 4, 20
Panasonic, 2, 19
Panasonic PK-700A video camera, 84
Parts replacement, 58
Pause button, 28-29
Perpendicular recording, 18
Phillips/Magnavox system, 126
Pickup tube comparison, 90
Pickup tubes, 79
Picture quality, 8
Pictures to electric impulses, converting, 17
Pictures with light streaks, 90
Piracy, 58, 102-103, 107
Piracy, tape, 124
Pirate decoder manufacturers, 123-124
Pirate decoders, 72
Playback, 16
Playback head, 16
Play button, 27, 29
Plumbicon, 90
Portable recorders, 60
Power consumption, 81
Principles, basic, 14
Product selection, 50
Program audio, 78
Projection, high-quality, 98
Projectors, 98
Public domain, material in, 124
PVC, 19

Q

Quadruplex recording, 5, 18, 20
Quasar "Time Machine," 23

R

Radio Shack, 67
RCA, 2, 10, 82
RCA Selectavision, 126
RCA TK-76 video camera, 75, 112
RCA VBT-200 recorder, 27
Receiver, 25
Record button, 29
Record function, activating the, 27
Record head, 14
Record heads, viii
Record/playback deck, 40
Recorder, industrial and educational, 25
Recorder, reel-to-reel, 2
Recorders, portable, 60

Recording, magnetic, 15
Recording at fast speed, 59
Recording process, 14
Recording tape, 16
Recording time, 50
RED-EO-TAPE production unit, 6
Registration, 85
Remote control, 51, 78
Repairs, expensive, 57
Resolution, 83, 88
RETMA Standard Resolution Chart 9, 85
Retrieval process, 14
Return signal, 77
Return video, 77
Rewind button, 27, 29

S

Sanyo, 82
Sanyo portable recorder, 47
Sanyo-V-Cord recorder, 23
Satellite telecasting, 74
Satellite TV, 123
Saticon, 60, 90
Saticons, broadcast quality, 79
Scan, helical, 7, 19
Scanner, 18, 36
Scanner speed, 36
Scan rate, 5, 7, 83
Scan rate difference, 69
Scenes, slating, 118
Script development, 110
Scripting, 112, 136
SECAM system, 4, 20
Selectavision, RCA, 126
Sensing system, 32, 34
Sensors, detector, 32
Service availability, 51
Service organization, choosing a, 52
Service training, 51
Servicing VHS equipment, 35, 51
Servo circuits, 45
Shooting, trial, 113
Shooting on location, 113
Shots, 113
Signal-processing package, 44
Signal source subassembly, 40
Signal standards, 4
Slant track recording, 19-20
Slow-scan television (SSTV), 4
Software for disc players, 127
Sony AV-3000 recorder, 1
Sony BVH-500 recorder, 111
Sony Corporation, 1, 19
Sony KV-1920, 25
Sony SL-7200 recorder, 2, 27, 36

Sony SL-8200 recorder, 22
Sony SLO-320, 28
Sony SLO-320 recorder, 25, 29
Sony Standard Test Tape, 25
Sony 2800 recorder, 25
Sony VO-1600 recorder, 2
Sony VO-1800 recorder, 21
Sound, stereophonic, 48, 67
Sound to tape, adding, 28
Splicing, 66
Sports finder, 80
SSTV, 4
Standards, industry, 20, 52, 73
Station selector button, 31
Stereophonic sound, 48, 67
Stop button, 27, 29
Story line, 110, 131
Studio finder, 80
Subscription TV, premium, 123
Sync, 113
Sync, provisions for external, 77
Synchronizing signals, 8
Sync signal, 76
System control unit, 41

T

Tally, 78
Tape, 102
Tape, best consumer grade, 106
Tape, grades of, 104
Tape, magnetic, viii
Tape, pre-recorded, 102
Tape, recording, 16
Tape, recording black over a, 119
Tape, recording pre-recorded, 58
Tape, sub-standard, 104
Tape duplication, 108
Tape for difference situations, choosing, 105
Tape generation, 63
Tape handling, 25, 32
Tape lifespan, 106
Tape motion, 32
Tape path, 33
Tape piracy, 102-103, 107, 124
Tape rental, 102
Tapes, bad, 56-57
Tapes, best, 58
Tapes, cassette-encased, reel-to-reel cartridge, 32
Tapes, longer playing, 22
Tapes, pre-recorded, 106
Tapes, purchasing pre-recorded, 107
Tapes, renting, 107
Tape size, 11
Tape speed, 18
Tape threading, 25

Tape transport, 40
Tape writing speed, 19, 25
Tele-Cine film chain, 69
Tele-Cine slide chain, constructing
    your own, 70
Time base correction, 59
Time-base error, 53
Time code, laying the, 118
Time Code system, 28
Timer, 41
Time-shift recording, 4
Tilting device, 2
Tint, color, 100
Tracking knob, 31
Track width, 36, 39
Transition editors, 121
Tuner, checking your TV, 55
Tuner, TV, 55
Tuner, vhf./uhf, 21-22
TV, checking your, 55
TV, CTR, 97
TV, large screen, 97
TV, solidstate, 94-95
TV receiver, 47, 53, 94
TV tuning, 95
Type C format, 2, 20, 111

VHS format, 2, 20, 22, 29, 58
VHS tape handling system, 33
Video and audio output jacks, 8
Video cameras, 60, 75
Video cassette recording (VCR), 40
Video clubs, 102
Video disc players, 126
"Video enhancer," 63
Video head assembly, 36
Video head cleaning, 58
Video Home System (VHS), 2
Video signals, systems for transmit-
    ting, 4
Videotape in business, viii
Videotape in education, vii
Videotape in homes, viii
Videotape in law, ix
Videotape recorder, first, 12
Videotape splicing, 66
Vidicon, 90
Vidicon tube, 17
Viewfinder, 80, 84
Viewfinders, electronic versus optical,
    89
Voice over, 113
VTR, viii

U

U-Matic cassette tape sizes, 106
U-Matic format, 2, 19-20, 25

V

Varactor tuning, 31, 95
VCC installation, 31
VCR, dismantling a, 57
VCRs, home versus industrial, 59
VHD/AHD system, 126-127

W

Walt Disney Productions, 122
Warnings, 26, 57, 66, 72, 107
Warranty, 51
Wave diagrams, 15
World Radio & TV Handbook, 5

Z

Zero-guard-band recording, 37
Zoom lens, 81, 90